RICHARD RICE

THE AI GROWTH CODE FOR BUSINESS LEADERS

Lessons from 126 Companies Winning with AI

V inValuable

First Edition: January 2025

ISBN (paperback): 979-8-9924154-1-4
ISBN (hardcover): 979-8-9924154-4-5
ISBN (digital): 979-8-9924154-2-1

BISAC COM100000 COMPUTERS / Artificial Intelligence / Generative AI

CONTENTS

Foreword

FOREWORD

Right now, your competitors are using AI to cut costs, steal customers, and outperform businesses that are stuck trying to figure out where to start. Companies using AI are seeing 20-40% increases in sales conversions and 25-45% improvements in customer engagement.

I've watched it happen dozens of times. While some business leaders navigate the maze of AI tools and conflicting advice, others are already using AI to transform their businesses.

Here's the reality I've learned from implementing AI in over 120 companies: You don't need to waste months researching options or risk hundreds of thousands on over-complicated solutions. There's a clear path forward, and it's simpler than you might think.

I've spent twenty-five years watching businesses struggle with overpriced technology solutions, over-complicated implementations, and underwhelming results. As a Chief Marketing Officer and, more recently, a Chief AI Officer, I've seen firsthand how the gap between AI hype and business reality is costing companies like yours real money and market share.

That stops now.

What makes this book different is that you won't just hear from me; **you'll hear directly from business leaders just like you**. I've spent hundreds of hours analyzing meeting transcripts, emails, and implementation notes to bring you the authentic voices of business leaders who've successfully navigated this journey. No theory, no jargon, no false promises, just real strategies that deliver real results for businesses like yours.

With the AI GROWTH Code, developed and refined through real-world implementations, I'll show you:

- The reality of what AI can (and can't) do for your business
- Your personal road map to AI implementation success
- Actual results from companies who've succeeded with AI.

You don't need a massive budget or a team of data scientists. You need a proven approach that works for businesses your size. That's exactly what you'll find in these pages.

Every strategy, every case study, and every tool comes from direct experience.

- You'll learn how Nina B. at Green Valley Landscaping cut fuel costs by 30% with AI route optimization after previously wasting $4,000 monthly on inefficient routes and spending 15+ hours weekly on manual scheduling.
- How Tom at EndUp Furniture reduced late payments by 50% and increased cash flow by 35% after struggling with a third of their accounts being overdue and spending three hours daily chasing payments.
- How Olivia at Bon Secours Hospital slashed hiring time in half while improving retention by 13%, after previously taking 45 days to fill critical positions and wasting $65,000 annually on inefficient recruitment processes.

These aren't just case studies; they're your future colleagues in AI success, sharing their challenges, breakthroughs, and victories.

By the time you finish this book, you'll have:

- *Clarity about AI's real potential for your business*
- *Confidence in choosing the right solutions*
- *A practical plan for implementation*

All without falling for the hype or getting bogged down by technical details.

Waiting months to start your AI journey in today's market could put you years behind your competitors.

The AI future isn't coming—it's here.

And it belongs to businesses that leverage AI effectively.

Let's make sure yours is one of them.

Richard F. Rice II
Founder, The AI Pros Agency
Creator, The AI Operating System™
Keynote Speaker, AI Pro Summit

Throughout this book, you'll hear from business owners just like you who faced this exact moment of decision. Some waited, watching competitors pull ahead. Others took action, following a proven path to success. Their stories, struggles, and triumphs will help guide your journey.

Let's begin by understanding exactly where your business stands in today's AI landscape.

PREFACE

THE REAL STORY OF AI IN SMALL BUSINESS

Right now, thousands of businesses are facing a critical choice: Rush into AI without a clear strategy and risk costly mistakes, or wait too long and watch competitors pull ahead.

In the past three years alone, I've seen companies transform routine operations into strategic advantages.

- Teams working smarter, not harder.
- Customer satisfaction soaring.
- Market share growing.

But I've also watched businesses waste hundreds of thousands on unnecessary solutions or lose market position while stuck in "analysis paralysis."

The difference between success and failure isn't about having the biggest budget or the most advanced technology.

It's about having the right approach.

THE REALITY CHECK FRAMEWORK

Throughout this book, you'll find Reality Check sections that look like this:

These Reality Checks serve as your guide to separating AI fact from fiction. They're based on real situations I've encountered and questions I've been asked while helping businesses like yours.

✓ REALITY CHECK

Myth: AI requires massive budgets and technical expertise.

Reality: Small businesses can implement effective AI solutions for as little as $20/month.

Impact: Many SMBs miss opportunities by assuming AI is out of reach.

AI Pro Tip: Start with basic AI tools that integrate with existing systems.

This book is structured around three essential elements to your AI success. Each part builds on the previous one, giving you a clear path forward:

PART I:
DECODING AI'S BUSINESS REALITY

The first part of our journey tackles the real state of AI for businesses like yours. No hype, no fear-mongering, just straight talk about what AI can and can't do. You'll learn:

- How to identify your highest-impact AI opportunities
- Which solutions fit your budget and business model

Preface

- Where to find quick wins that build momentum
- Common pitfalls and how to avoid them

PART II:
THE AI GROWTH CODE

Next, I'll share my proven GROWTH code, the same system that's helped hundreds of businesses succeed with AI:

- **G**oals: Turn business challenges into clear AI objectives
- **R**esources: Build your AI capability without breaking the bank
- **O**perationalize: Make AI work in your daily operations
- **W**iden: Expand your success across the organization
- **T**ransform: Revolutionize your business processes
- **H**arness: Maximize and sustain your AI advantage

PART III:
EXECUTING THE CODE

Finally, we'll explore the practical steps to start your AI journey and ensure lasting success. You'll learn:

- How to begin your AI implementation with confidence
- How to measure and maximize your return on AI investment
- How to future-proof your business in an AI-driven world

Every chapter in this section builds on the GROWTH Code and strategies we've covered, showing you exactly how to put them into action in your business.

THE AI GROWTH CODE TOOLKIT

To help you implement each step of the GROWTH Code successfully, this book includes more than a dozen battle-tested tools, templates, and frameworks worth over $50,000 in consulting value. Each tool has been refined through successful implementations at over 120 businesses, saving you months of trial and error and thousands in consulting fees.

These practical resources are strategically placed throughout the book to support your journey:

- Part I includes readiness assessments and planning tools
- Part II provides implementation templates and frameworks for each GROWTH component
- Part III delivers execution guides and measurement tools

Every tool can be accessed online through aigrowthcode.com, making it easy to implement what you learn immediately.

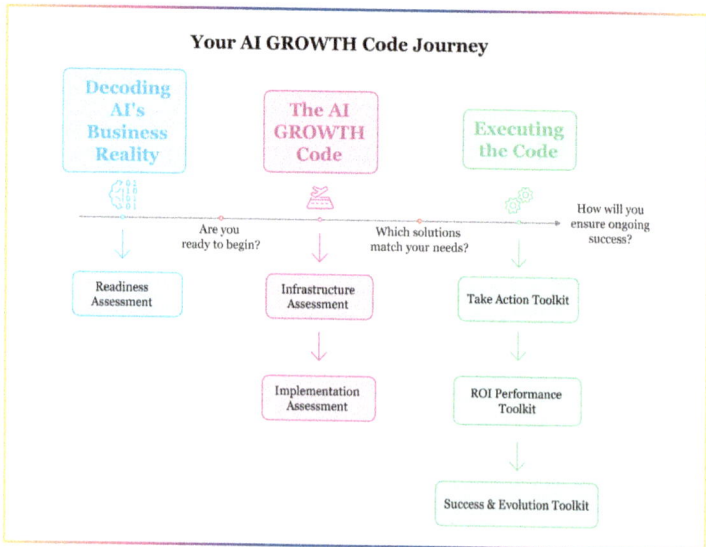

Your AI GROWTH Code Journey

Decoding AI's Business Reality — Are you ready to begin? — Readiness Assessment

The AI GROWTH Code — Which solutions match your needs? — Infrastructure Assessment — Implementation Assessment

Executing the Code — How will you ensure ongoing success? — Take Action Toolkit — ROI Performance Toolkit — Success & Evolution Toolkit

The journey of a thousand miles begins with a single step — but that step should be in the right direction.

I'm personally committed to making sure yours is pointed toward success.

That commitment comes from seeing too many businesses learn expensive lessons about AI implementation. Tallwave Marketing recently called me after spending six months and $43,000 trying to build their own AI social media management system. Their tech-savvy digital manager had convinced them they could piece together a solution using free tools and YouTube tutorials. By the time they reached out, they were spending more time managing disconnected AI tools than creating meaningful engagement across their six social platforms.

Within three weeks of implementing the right approach, their team saved 10 hours weekly and increased follower engagement by 12%. The digital manager admitted, *'We thought having technical knowledge was enough. We didn't know what we didn't know.'*

That success starts with understanding exactly how AI is reshaping business today. The landscape has shifted dramatically, and knowing where you stand is the first step to knowing where you can go.

DECODING AI'S BUSINESS REALITY

Now, let's cut through the confusion and get real about AI for your business. There are no empty promises or theoretical concepts; there are just practical insights from businesses that have already successfully taken this journey.

In this section, you'll learn:

- The truth about what AI can (and can't) do for companies today
- Where other companies are seeing the biggest returns
- Which AI myths are costing businesses time and money

Part One: Decoding AI's Business Reality

CRACKING THE CODE HOW AI IS SHAKING UP BUSINESS

"There are going to be two kinds of companies by the end of this decade: those that are fully utilizing AI and those that are out of business."

PETER DIAMANDIS-FUTURIST [1]

THE AI REVOLUTION IN SMALL BUSINESS: BY THE NUMBERS

The business landscape is experiencing a seismic shift. AI adoption among small and medium-sized businesses more than doubled in just one year, jumping from 14% to 39[2] between 2023 and 2024. This isn't just a trend—it's a transformation backed by compelling financial results.

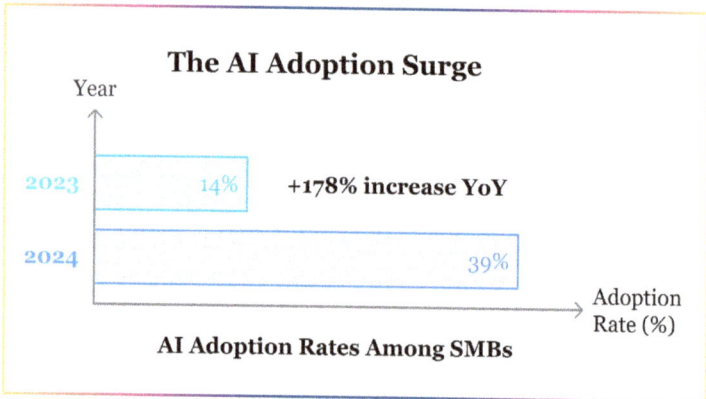

The AI Adoption Surge

Year

2023   14%   **+178% increase YoY**

2024   39%

Adoption Rate (%)

AI Adoption Rates Among SMBs

THE NUMBERS THAT MATTER

Recent studies clearly show AI's impact: Companies implementing AI solutions see an average 15.2% reduction in costs and a 15.8% increase in revenue [3].

A Forrester study projects that businesses implementing AI solutions will achieve ROIs ranging from 132% to 353% over three years [5].

McKinsey's research reveals a stark divide in the business world: 59% of companies that properly implement AI see increased revenue, while those without systematic implementation struggle to see results. Among successful implementers, 42% have also achieved significant operational cost reductions [3].

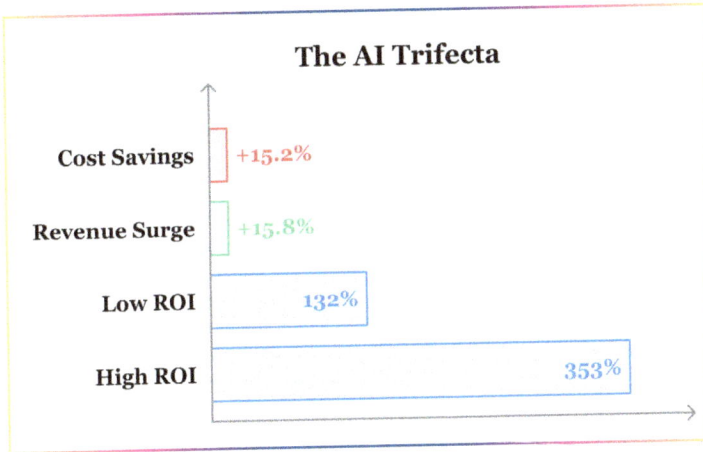

The AI Trifecta

Cost Savings	+15.2%
Revenue Surge	+15.8%
Low ROI	132%
High ROI	353%

MICRO-BUSINESS SUCCESS METRICS

When Laura's Flower Shop (3 employees, $380K annual revenue) started using basic AI tools:

- Initial investment: $240/year
- Reduced order processing from 15 to 2 minutes
- Cut flower waste by 40%
- ROI: $48,000 annual revenue increase

"We thought AI was only for big companies," Laura shared. *"But with just basic tools, we've transformed how we handle orders and manage inventory."*

✓ REALITY CHECK

Myth: AI success requires significant business scale.

Reality: Even micro-businesses achieve substantial returns with minimal investment.

Impact: Small businesses often delay adoption, thinking they're too small.

AI Pro Tip: Start with one specific business problem that wastes time or money daily.

THE RACE IS ON

But here's what should keep you up at night: While reading this, 31% of your competitors are developing their AI strategy [4]. They're not just planning—they're acting. Add them to the 39% who have already adopted AI in some form, and suddenly, 7 out of 10 of your competitors are leaving you behind.

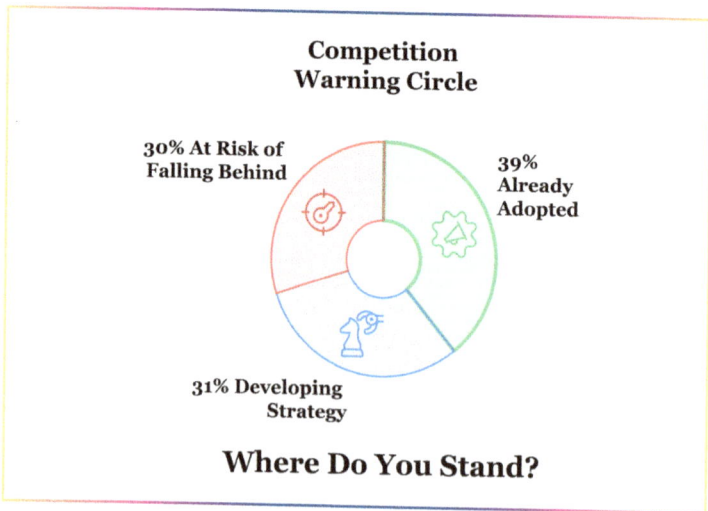

Competition Warning Circle

30% At Risk of Falling Behind

39% Already Adopted

31% Developing Strategy

Where Do You Stand?

The message is clear: The question is not whether to adopt AI but when and timing matters.

91% of SMBs believe AI will help their business grow [3]. But only 39% have taken action [4]. This gap between belief and action represents both a challenge and an opportunity for you.

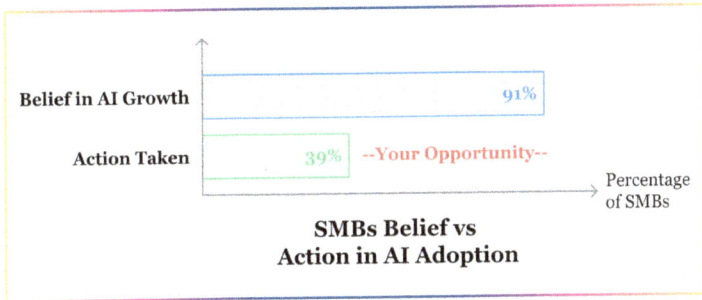

SMBs Belief vs Action in AI Adoption

THE EXPERTISE GAP

The most sobering statistic is that the main barrier to AI adoption isn't technology or Cost—it's expertise. 37% of businesses cite a lack of expertise as their primary obstacle [4]. This is exactly why I've spent the last three years helping businesses bridge this gap and why I wrote this book.

Let me show you how to join the 39% of businesses that are already using AI to boost their revenue [6] rather than watching from the sidelines as your competitors gain ground.

BEYOND THE NUMBERS

These aren't just numbers on a page—they're the reality I see unfolding every day with clients like Manuel G., a seasoned accounting firm partner managing 16 employees and $4.2M in annual revenue. When I first sat down with him, he admitted something I often hear: *"I thought AI was rocket science. Turns out, I was making it more complicated than it needed to be."* After implementing basic AI tools, his firm reduced data entry errors from 4% to 0.1%, saving $125,000 annually in rework and client satisfaction issues.

The secret? Understanding what AI is...and isn't.

MAKING AI WORK FOR YOU

In the past three years, I've helped more than 120 businesses successfully implement AI into their operations. During that time, I've learned that success with AI isn't about chasing the latest buzzwords or most advanced technology.

It's about solving real business problems with practical solutions. Whether you're running a local retail store, a healthcare practice, or a manufacturing facility, the principles remain the same:

- Start with clear business problems,
- Implement proven solutions,
- Build on your successes.

In this chapter, I'll show you exactly how businesses like yours use AI to solve everyday challenges and why now is critical to begin your AI journey.

PRACTICAL ROI EXAMPLE: SWEET SUCCESS BAKERY

Mary's downtown bakery wasted ingredients and labor hours by baking too much on slow days and running out on busy days. Without an effective way to predict daily demand, it lost about $2,000 monthly in wasted products and missed approximately $3,000 in potential sales.

They implemented a basic AI system that:

- Analyzed historical sales data
- Considered local events and weather
- Predicted daily item demand
- Adjusted production schedules
- Generated ingredient orders

Results after 90 days:

- Initial Investment: $600
- Monthly Cost: $50
- Time Savings: 15 hours/week on planning and ordering
- Labor Value: $25/hour

Monthly Return:

- Time Value: $1,500 (reduced planning/ordering hours)
- Reduced Waste: $800 (40% reduction in expired items)
- Increased Sales: $2,000 (better product availability)
- Net Monthly ROI: $4,250
- Break-even: 5 days

THE AI TIDAL WAVE

I often get asked, *"Is AI really different from all the other tech trends we've seen?"* Let me share what I've observed across many industries: Unlike previous technological shifts that changed how we do things, AI changes our fundamental problem-solving capacity.

Think about how the internet changed everything in the '90s and early 2000s. It connected us globally, but we still had to figure out what to do with those connections.

AI goes further – it doesn't just connect information; it helps us understand it and act on it.

"AI is the new electricity. Just as electricity transformed almost everything 100 years ago, today I have a hard time imagining an industry that will not be transformed by AI in the next several years"

ANDREW NG – CO-FOUNDER AT COURSERA

✓ REALITY CHECK

Myth: AI is only for tech companies or large corporations.

Reality: Businesses of all sizes and industries are successfully using AI.

Impact: Delayed adoption puts SMBs at a competitive disadvantage.

AI Pro Tip: Start with basic AI applications that solve

SPECIFIC BUSINESS PROBLEMS

A NEW ERA OF AUTOMATION

In my 25 years of helping businesses adopt new technology, I've never seen anything transform operations as quickly as AI.

I've watched businesses repeatedly achieve in weeks what used to take months or years of process improvement.

But here's what makes AI different: It does not just automate tasks— it enhances decision-making at every level.

When businesses implement AI effectively, they don't just save time on routine tasks. They discover patterns and insights that help them prevent problems before they happen.

That's the fundamental shift I see with AI: it turns **data into insights and insights into action.**

REAL-WORLD SUCCESS METRICS

Across a wide variety of industries, here's what I consistently see:

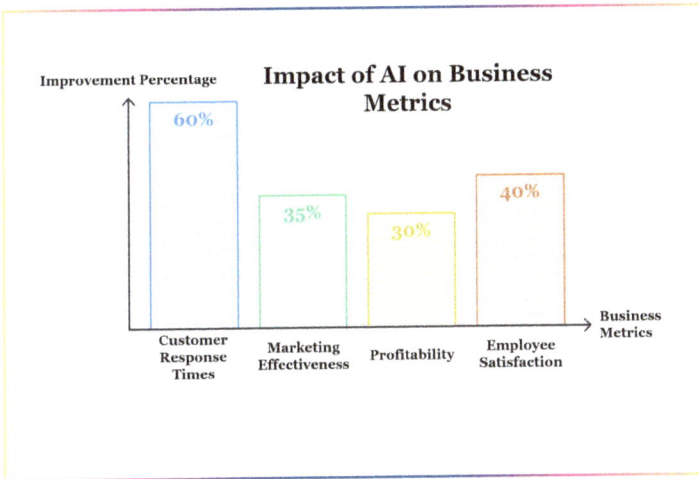

Impact of AI on Business Metrics

A recent McKinsey Quarterly study found that AI can reduce operational costs by up to 30% through automation and streamlined processes. [7}

These results aren't cherry-picked; they represent the typical outcomes of implementing AI correctly.

WHY SMALL BUSINESSES CAN'T IGNORE AI ANYMORE

Three key factors make this a critical moment for SMB AI adoption:

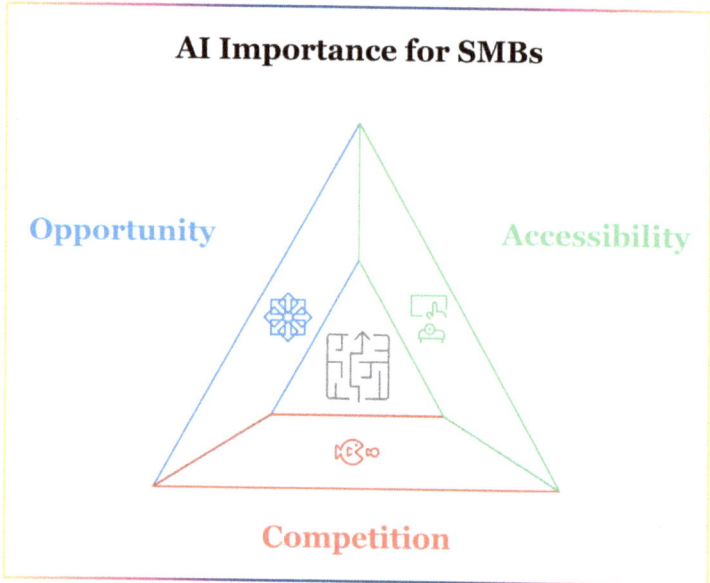

AI Importance for SMBs

Opportunity

Accessibility

Competition

Early adopters of AI can leverage significant market advantages:

- First mover benefits in customer service
- Competitive edge in operational efficiency
- Time to learn and optimize before competitors

AI tools are now more affordable and user-friendly, making them accessible to SMBs.

- Basic AI solutions starting at $0-1,200/year
- User-friendly implementations
- Short setup times (1-4 weeks)

(Much more detailed pricing in Chapter 8)

Larger companies use AI to gain a competitive advantage targeting SMB customers:

- Advanced customer analytics
- Personalized marketing at scale
- Data-driven competitive strategies

Throughout this book, we will cover case studies that show how businesses achieve these advantages.

✓ REALITY CHECK

Myth: AI implementation requires massive investment.

Reality: Many SMBs achieve significant ROI with basic AI tools.

Impact: Fear of costs prevents businesses from capturing available benefits.

AI Pro Tip: Start with targeted solutions that address specific pain points.

WHY YOU CAN'T AFFORD TO WAIT

"But Rich," you might be thinking, *"my business is doing fine without AI."* Sure, maybe it is, for now. But remember Blockbuster? They were doing fine, too, until Netflix came along with its data-driven, personalized approach. Change is speeding up, and AI is the driving force.

Integrating AI into your business isn't just about staying relevant; it's about preparing to thrive in a future where AI is as fundamental as electricity. It's about making better decisions faster, understanding your customers more deeply, and innovating in ways you might not have imagined.

BREAKING DOWN AI BARRIERS

I've noticed these common challenges with my SMB clients when adopting AI. They're crucial to address if you want AI actually to benefit your business:

1. Cost and Resources:

AI isn't cheap. It requires both money and talent, and for most SMBs, those are in short supply. Hiring AI specialists or investing in AI tools can feel out of reach when you're already running lean.

2. Data Issues:

AI thrives on data, and SMBs don't have the huge datasets that bigger companies have. Even when they collect data, it's often not clean or structured enough to be useful for AI. Without good data, AI is basically running blind.

3. Lack of AI Expertise:

Most SMBs don't have data scientists or AI specialists on staff. This means they're relying on third-party solutions, which can be confusing and overwhelming. AI isn't plug-and-play; it's tough to implement effectively without the right people.

4. Integration with Legacy Systems:

Many SMBs rely on older systems that aren't compatible with AI tools.

5. Misunderstanding AI:

Many SMB owners think AI is a magic bullet. It's not. I've seen too many clients expect AI to fix all their problems overnight without realizing that it is a tool, not a cure-all. You need clear goals and realistic expectations.

6. Ethical and Compliance Concerns:

AI raises many red flags regarding data privacy and security. Many SMBs lack the legal teams or compliance experts needed to navigate these issues, leaving them open to serious risks.

7. Scalability:

Most AI solutions are built for large companies, making them too complex or expensive for smaller businesses. Finding the right-sized solution is often a struggle.

8. Internal Resistance:

Employees and leadership can sometimes push back against AI adoption, usually out of fear that it'll replace jobs or because they don't understand its value. That resistance can stall progress before you even get started.

THE TRUE POWER OF AI

Now that we've discussed the challenges, let's examine how AI can help your business grow. AI isn't just for big companies; it can also give SMBs like yours a real boost.

1. Work Smarter, Not Harder

AI can handle repetitive tasks, freeing up your and your team's time. Imagine a helper that can:

- Sort through emails and answer common questions
- Keep track of inventory automatically
- Schedule appointments without playing phone tag

This means you can focus on important things like growing your business and making customers happy.

2. Know Your Customers Better

AI helps you understand what your customers want, sometimes even before they do. It can:

- Suggest products customers are likely to love
- Predict when a customer might need help
- Personalize marketing to each customer's interests

Happy customers mean a healthier business.

3. Make Smarter Decisions

AI is like having a super-smart advisor. It can:

- Spot trends in your sales data
- Predict busy seasons so you can plan ahead
- Help you decide where to open a new location

With AI, you make choices based on solid information, not just gut feelings.

4. Come Up with New Ideas

AI can spark creativity in your business. It might:

- Suggest new product features based on customer feedback
- Help design eye-catching ads
- Find new ways to use your existing resources

✓ REALITY CHECK

Myth: AI will replace human decision-making.

Reality: AI enhances human capabilities rather than replacing them.

Impact: Businesses that view AI as a replacement tool miss its true value.

AI Pro Tip: Focus on how AI can augment your team's existing capabilities.

Let me show you how these principles work in practice. When I first met Jane at Mason & Magnolia Décor, she was skeptical about how AI could help a traditional home décor store. A family-owned business with $2.8M in annual revenue and 12 employees, it faced growing pressure from online competitors. Her journey from skeptic to advocate demonstrates exactly how these benefits translate into real business results.

CASE STUDY: MASON & MAGNOLIA DECOR

INITIAL SITUATION

- 2nd generation family-owned home décor store
- 12 employees struggling with customer response times
- Customer queries going unanswered for hours
- 3.5/5 star rating due to slow response times
- Losing business to online competitors
- Limited staff availability after-hours

IMPLEMENTATION JOURNEY

Started with basic AI

- Implemented ChatBot Pro, trained on the store's product data
- Automated response system for common questions
- 24/7 customer support capability

Added inventory management

- Real-time stock updates
- Automated reordering
- Demand forecasting

Integrated marketing automation

- Personalized product recommendations
- Automated email campaigns
- Customer behavior analysis

Expanded to predictive ordering

- Seasonal trend analysis
- Supply chain optimization
- Inventory forecasting

RESULTS:

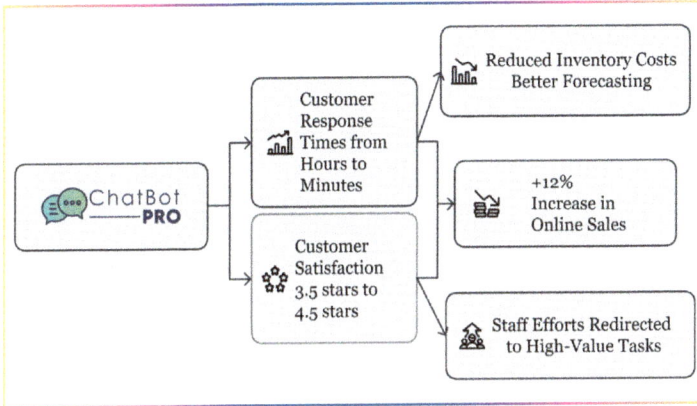

"Since implementing ChatBot Pro, customer service response times have decreased from hours to minutes, which increased our customer satisfaction ratings from 3.5 to 4.5 stars and increased online sales by 12%."

– JANE T., OWNER, MASON & MAGNOLIA DÉCOR

SUPPORTING SUCCESS STORIES

An E-commerce BBQ Sauce seller achieved the following:

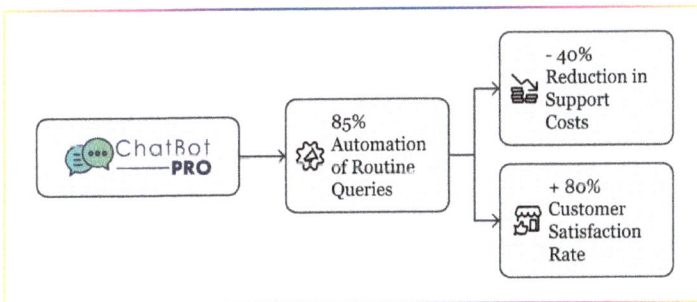

Chapter 1: Cracking the Code: How AI is Shaking Up Business

A Tech Support Manager in the EdTech space realized:

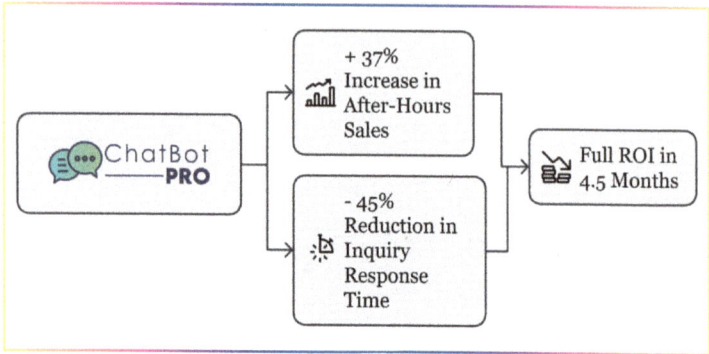

After seeing these results, the most common question I hear is: *"Rich, how do I know if my business is ready for AI?"* That's exactly why I developed the AI Readiness Assessment. Just as I helped Jane at Mason & Magnolia understand their starting point, this tool will help you:

- Identify your business's strongest opportunities for AI implementation
- Evaluate your current technological readiness
- Assess your team's capability for AI adoption
- Determine which AI solutions align with your resources
- Create a clear road map for moving forward

Assessment A - AI Readiness Assessment

Before diving deeper into implementation, it's crucial to understand where your organization stands today. That's why I've created the AI Readiness Assessment - a comprehensive evaluation that examines your business's strategic alignment, cultural preparedness, and change management capabilities. After completing this assessment, you'll receive

personalized recommendations and a clear action plan tailored to your situation. This assessment has helped organizations identify critical gaps early, often reducing wasted spending by 40% or more.

CHAPTER TAKEAWAYS

THE AI REVOLUTION IS HERE

- SMB adoption more than doubled in 2024 (14% to 39%)
- Companies seeing an average 15.2% cost reduction and 15.8% revenue increase
- ROI ranging from 132% to 353% over three years

THERE IS STILL A WINDOW OF OPPORTUNITY

- 41% of businesses developing AI strategy
- 31% believe in AI's potential, but only 29% take action
- First-mover advantage is still available in most industries

BREAKING THROUGH BARRIERS

- Expertise, not technology, is the main obstacle
- Success comes from systematic implementation
- Start with clear business problems, not technology solutions

LOOKING AHEAD

Being ready for AI isn't just about having the right technology; it's about having the right mindset. I've seen businesses with modest technical resources achieve remarkable results because they approached AI with the right perspective. I've also seen well-funded organizations struggle because they missed this crucial foundation.

Chapter 2: will explore how successful businesses developed this winning mindset and how you can do the same.

1. Diamandis, P. (May 17, 2022). https://x.com/PeterDiamandis/status/1526738108036927491
2. https://www.cpapracticeadvisor.com/2024/10/17/smbs-are-driving-growth-in-ai-adoption/111960/
3. https://blog.shi.com/digital-workplace/ai-is-not-just-for-giants-how-small-businesses-can-harness-its-power/
4. https://colorwhistle.com/artificial-intelligence-statistics-for-small-business/
5. https://www.microsoft.com/en-us/microsoft-365/blog/2024/10/17/microsoft-365-copilot-drove-up-to-353-roi-for-small-and-medium-businesses-new-study/
6. https://www.forbes.com/sites/larryenglish/2024/09/26/ais-roi-can-be-elusive-but-that-doesnt-mean-the-hype-is-misguided/
7. https://www.mckinsey.com/capabilities/quantumblack/our-insights/the-economics-of-artificial-intelligence

REWIRING YOUR BRAIN FOR AI SUCCESS

"The capacity to learn is a gift; the ability to learn is a skill; the willingness to learn is a choice."

BRIAN HERBERT, AMERICAN AUTHOR [1]

AI THINKING: A MINDSET REVOLUTION

Imagine walking into Justin's Family Market, a 12,000-square-foot grocery store that has served Milwaukee's east side for three generations. With 15 employees and $4.3M in annual revenue, it's a mid-sized independent grocer competing against major chains. As you browse the aisles, something stands out. The shelves are perfectly stocked, with just enough of each product, no empty spaces, and no overflows. You ask Justin about it.

"It's our new AI system," he beams. *"It predicts what we'll need based on weather forecasts, local events, and historical sales data. We've cut waste by 30% - about $156,000 annually - and increased profits by 20%, adding another $128,000 to our bottom line in just six months."*

What's truly revealing is what Justin says next, *and it surprises most people.* "*The technology was the easy part,*" he explains. "*The real change was in how we started thinking about our business. Once we stopped seeing inventory as something to manage and started seeing it as something to predict, everything changed. Now we're finding AI opportunities everywhere.*"

Not every business successfully makes this mindset shift. Let me share a costly lesson about the importance of mindset. Secured Tech Solutions called me after spending $78,000 on advanced AI support tools that were gathering digital dust. Their support team, who had managed education technology issues the same way for over two decades, insisted their existing processes worked fine. Despite having powerful new AI capabilities at their fingertips, the team kept manually sorting customer inquiries and handling device management the old way.

When executive management reached out, the support backlogs were worse than ever. The eight-person support team drowned in over 280 daily queries, and the costly AI investment sat unused. The team's resistance to change was costing more than money; it was costing opportunities.

Within a month of helping them transform their mindset and processes, they handled 85% of customer queries through AI, dramatically improving their response rate and customer satisfaction. Their support team finally had time to focus on complex technical issues requiring human expertise. '*We thought our traditional processes were the backbone of our success,*' Joshua, their Ed Tech Support Manager, admitted. '*Now we understand that being good at what we've always done was actually holding us back from being great at what we could do.*'

FROM MANAGEMENT TO PREDICTION

This isn't science fiction. It's happening right now in businesses worldwide. Justin's Store demonstrates a fundamental truth about AI transformation: success begins with a shift in thinking.

The key point?

The mindset shift came first. The technology followed.

THE NEW WAY FORWARD

Welcome to the era of AI-enhanced business. It's not just about implementing new technology but fundamentally changing how we think and operate. This chapter will explore the AI Mindset, a shift from managing to predicting, and how it can transform your business.

HOW TO IDENTIFY AI-READY PROCESSES AND PAIN POINTS

First, you must identify the right opportunities to transform your business with AI. Let me show you exactly how to do that.

Let's start by understanding what makes a process ready for AI enhancement. I've identified five key characteristics that signal an opportunity for AI transformation. I use these practical indicators with every client to spot their highest-impact AI opportunities.

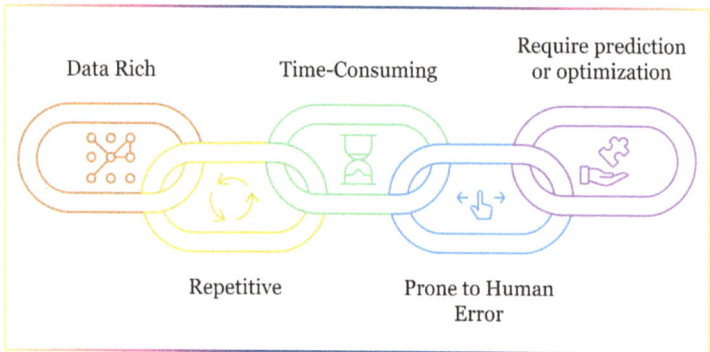

These characteristics consistently help businesses identify their best starting points for AI transformation. Let me give you a real-world example of how this works.

REAL-WORLD EXAMPLE

When Tom from EndUp Furniture first assessed his processes, he found their accounts receivable system checked all five boxes.

- Data Rich: Payment history, customer records, and seasonal patterns provided a wealth of information
- Repetitive: Regular payment processing and follow-ups consumed significant time
- Time-Consuming: The team spent 15+ hours weekly on collections
- Error-Prone: Manual payment tracking led to missed follow-ups
- Required Prediction: Cash flow forecasting was critical for business planning

Result. After implementing AI, they reduced late payments by 50% and improved cash flow by 35%.

Once you've identified these characteristics in your processes, the next step is asking the right questions.

THE ART OF ASKING THE RIGHT QUESTIONS

The difference between successful and struggling AI implementations often comes down to framing opportunities correctly. To uncover where AI could make a difference, start asking questions like:

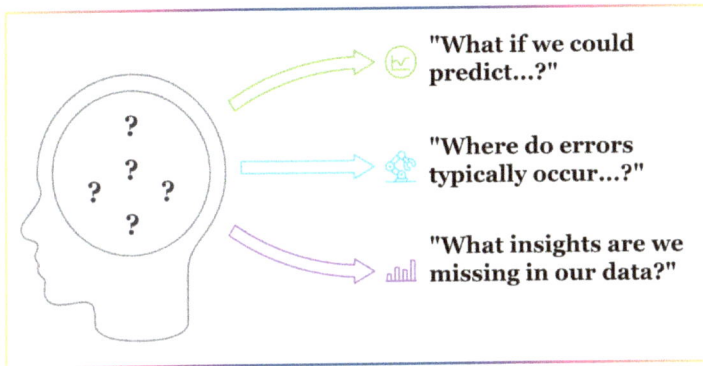

"What if we could predict...?"

"Where do errors typically occur...?"

"What insights are we missing in our data?"

Asking these types of questions shifts your approach from reactive problem-solving to proactive opportunity creation.

When business leaders start asking these strategic questions, they often experience a fundamental shift in their thinking.

This shift is about moving from reactive problem-solving to proactive opportunity creation.

Let's look at how this mindset shift transforms decision-making across multiple departments.

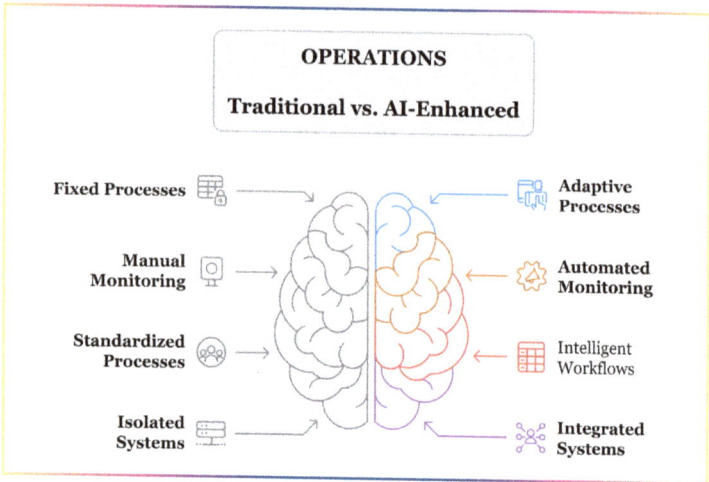

OPERATIONS

- *Traditional*: Reacting to equipment failures and process bottlenecks
- *AI-Enhanced*: Using predictive analytics to prevent issues and optimize workflows
- *Impact*: Like at Duke's Mayonnaise, where predictive maintenance reduced downtime by 75% and improved production efficiency by 30%

✓ REALITY CHECK

Myth: Operational AI requires a complete system overhaul.

Reality: Most businesses can start with basic predictive maintenance.

Impact: Starting small allows for quick wins and builds momentum.

AI Pro Tip: To prove value, begin with one critical process or machine.

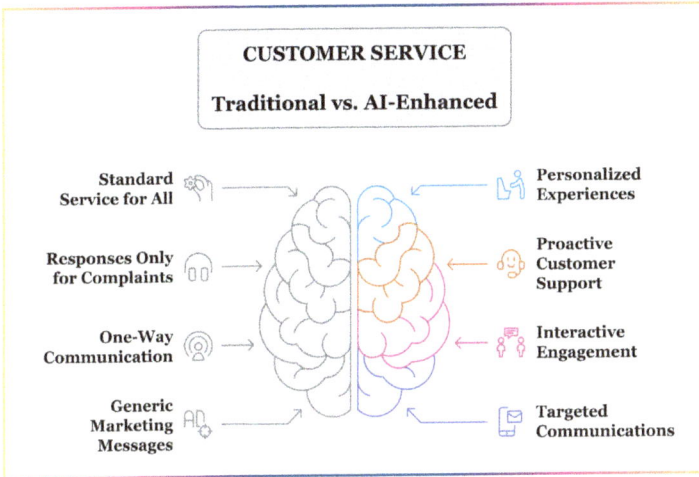

CUSTOMER SERVICE

Traditional vs. AI-Enhanced

Standard Service for All → Personalized Experiences

Responses Only for Complaints → Proactive Customer Support

One-Way Communication → Interactive Engagement

Generic Marketing Messages → Targeted Communications

CUSTOMER SERVICE

- *Traditional*: Waiting for customer complaints and questions
- *AI-Enhanced*: Proactively identifying and addressing potential issues
- *Impact*: As seen at Mason & Magnolia, where AI reduced response times from hours to minutes and improved satisfaction ratings from 3.5 to 4.5 stars in 6 short months

✓ REALITY CHECK

Myth: AI chatbots will frustrate customers.

Reality: Well-implemented AI improves response times and satisfaction.

Impact: Poor implementation can damage customer relationships.

AI Pro Tip: Start with common queries and expand based on customer feedback.

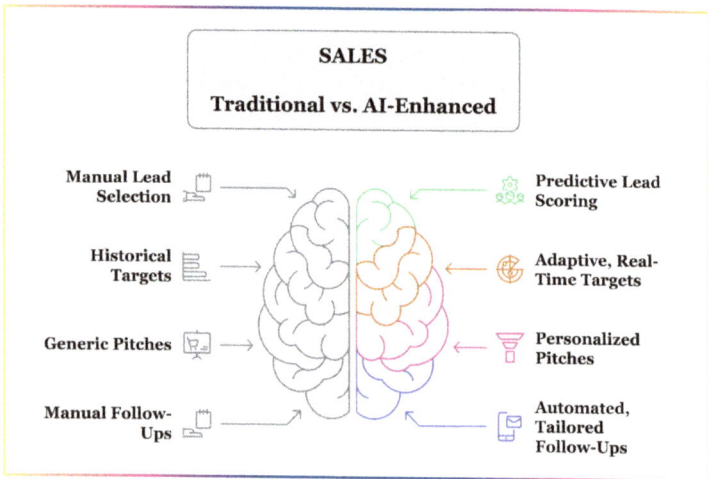

SALES

Traditional vs. AI-Enhanced

Manual Lead Selection	→	Predictive Lead Scoring
Historical Targets	→	Adaptive, Real-Time Targets
Generic Pitches	→	Personalized Pitches
Manual Follow-Ups	→	Automated, Tailored Follow-Ups

SALES

- *Traditional*: Relying on historical data and gut instinct for forecasting
- *AI-Enhanced*: Using AI to predict customer behavior and personalize approaches
- *Impact*: Like at Harry's Razors, where predictive analytics reduced customer churn by 30% and saved $200K annually

✓ REALITY CHECK

Myth: AI eliminates the human element in sales.

Reality: AI enhances sales teams' ability to build relationships.

Impact: Sales teams without AI support fall behind competitors.

AI Pro Tip: Use AI to identify high-value opportunities for personal follow-up.

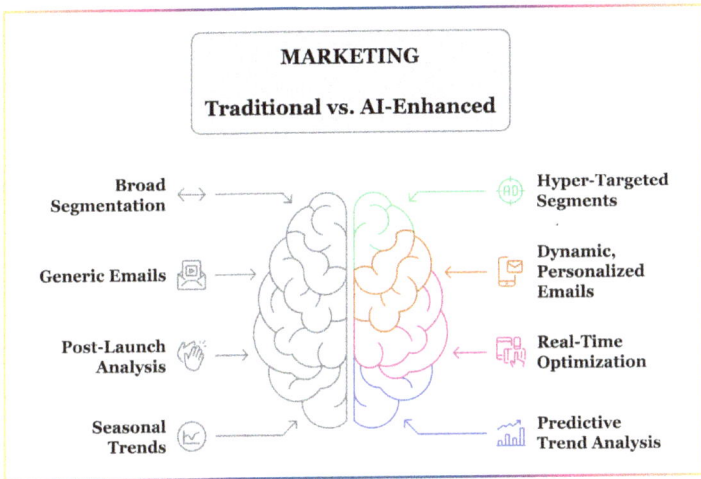

MARKETING

Traditional vs. AI-Enhanced

Broad Segmentation ⟷ Hyper-Targeted Segments

Generic Emails → Dynamic, Personalized Emails

Post-Launch Analysis → Real-Time Optimization

Seasonal Trends → Predictive Trend Analysis

MARKETING

- *Traditional*: Broadcasting generic messages to broad audiences
- *AI-Enhanced*: Delivering personalized content based on behavioral analysis
- *Impact*: As demonstrated at Tallwave Marketing, where AI-driven personalization increased engagement by 33% and saved 10 hours weekly

✓ REALITY CHECK

Myth: AI marketing lacks personalization and creativity.

Reality: AI enables deeper personalization while freeing marketers for creative work.

Impact: Generic messaging increasingly fails to engage customers.

AI Pro Tip: Use AI to test and refine messaging before the full campaign launch.

Chapter 2: Rewiring Your Brain for AI Success

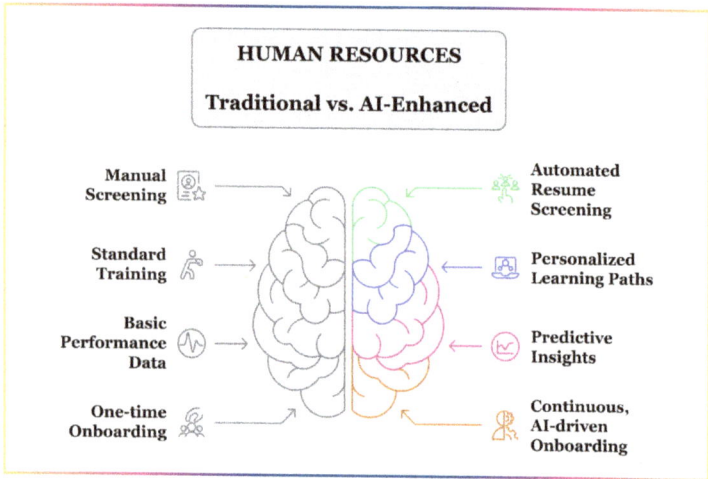

HUMAN RESOURCES

- *Traditional*: Manual resume screening and reactive hiring
- *AI-Enhanced*: Predictive talent acquisition and proactive retention strategies
- *Impact*: Like at Bon Secours Hospital, where AI reduced hiring time by 50% and improved retention by 13%

✓ REALITY CHECK

Myth: AI hiring eliminates human judgment.

Reality: AI screening enhances recruiters' ability to find the best candidates.

Impact: Manual screening misses qualified candidates and introduces bias.

AI Pro Tip: Start with basic resume screening before expanding to predictive hiring.

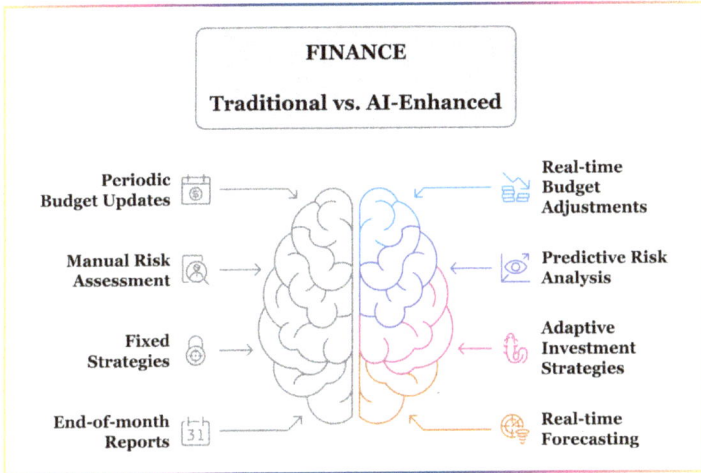

FINANCE

Traditional vs. AI-Enhanced

Traditional	AI-Enhanced
Periodic Budget Updates	Real-time Budget Adjustments
Manual Risk Assessment	Predictive Risk Analysis
Fixed Strategies	Adaptive Investment Strategies
End-of-month Reports	Real-time Forecasting

FINANCIAL MANAGEMENT

- *Traditional*: Monthly reporting and reactive cash flow management
- *AI-Enhanced*: Real-time financial insights and predictive forecasting
- *Impact*: As seen at EndUp Furniture, where AI-driven financial management reduced late payments by 50% and improved cash flow by 35%

✓ REALITY CHECK

Myth: AI financial decisions are too risky.

Reality: AI augments financial expertise with real-time data analysis.

Impact: Delayed insights cost businesses significant opportunities.

AI Pro Tip: Begin with cash flow forecasting to demonstrate reliable results.

WHY CHANGE YOUR APPROACH?

This shift in thinking from reactive to predictive isn't just about adopting new technology. It's about fundamentally re-imagining how each department can operate. The results consistently show that organizations embracing this mindset shift achieve significantly better outcomes than those implementing AI tools without changing their thinking.

Let me share a perfect example from Craig at TRAXX Flooring. In their first attempt at AI implementation, they automated their existing processes. "We spent $50,000 on automation," Craig told me, *"but we were just doing the same things faster."* Their results were modest, with a 15% efficiency improvement.

Then, they changed their approach. Instead of asking, *"How can we do this faster?"* they asked, *"Why are we doing this?"* This mindset shift led them to completely re-imagine their workflows. The result? A 42% reduction in process time and a 18% increase in customer satisfaction.

The numbers tell a clear story across industries:

- Companies that just implement AI tools: 15-25% improvement
- Companies that transform their thinking: 40-60% improvement
- Cost difference: Nearly identical investments

Myth: *Implementing AI tools alone will transform your business.*

Reality: *True transformation requires changing how you think about your operations.*

Impact: *Companies focusing only on tools achieve less than half the potential benefits.*

AI Pro Tip: *Start with the mindset shift, then choose your tools.*

TRUST YOUR GUT?

These transformative results often raise important questions among business leaders. Last month, I met with a successful manufacturing company owner who'd been in business for thirty years. Rich," he said, "*I've always trusted my gut. It's worked well so far. Why should I let AI make decisions for my business?*"

His question might be familiar to you. It's one of the most common concerns I hear from the business owners I work with. But here's what I told him and what I want to share with you:

AI isn't about replacing your business instincts; it's about enhancing them with data-driven insights that make your experience even more valuable.

✓ REALITY CHECK

Myth: *Traditional business models are safer and more reliable.*

Reality: *Traditional approaches often hide significant inefficiencies and missed opportunities.*

Impact: *Businesses lose competitive advantage while thinking they're playing it safe.*

AI Pro Tip: *Start identifying specific areas where traditional processes could be limiting growth.*

OVERCOMING LIMITATIONS OF TRADITIONAL BUSINESS MODELS

Traditional business models often have hidden inefficiencies. During a recent consulting session, Alex N., who owns a successful home services company with 45 employees and $21M in annual revenue, shared something revealing: *"I was so focused on maintaining our traditional processes that I couldn't see how much they were actually holding us back."* His team spent 15 hours per week sorting customer inquiries manually - a hidden cost of $42,000 annually in labor alone. After automating with basic AI tools, they reduced manual work by 95%, freeing up $40,000 worth of time for growth activities.

This is a common revelation among my clients. While familiar and comfortable, traditional business models often come with hidden costs that become apparent only when we examine them closely. Let me break down what I typically see:

TIME INEFFICIENCIES

Remember the days of paper filing systems? Most businesses have moved past those, but many still cling to equally outdated processes in other areas. For example, a distribution client spent $78,000 annually in labor costs just by having their team manually sort and route packages. After shifting their mindset and implementing basic AI tools, they automated 95% of this task, reclaiming $74,000 worth of productive time.

HIDDEN ERROR RATES

Human error is natural, but its cost isn't always visible. Take Alex's company. They thought their 4% error rate in data entry was "industry standard." Those errors silently cost them $156,000 annually in rework and customer satisfaction issues. After embracing an AI mindset and implementing simple automation, they reduced errors to 0.1%, saving $152,000 annually.

MISSED OPPORTUNITIES

Traditional approaches often involve playing defense rather than offense. One retail client manually analyzed customer purchase data, missing patterns that could drive sales. After implementing AI analytics, they discovered timing patterns that led to a $234,000 increase in annual revenue through better inventory management and targeted promotions.

Understanding these limitations is crucial, but it's only the first step. The real challenge and opportunity lies in transforming your organization's culture to embrace AI-enhanced thinking.

CULTURAL TRANSFORMATION IN ACTION

"Culture eats strategy for breakfast," management guru Peter Drucker famously said, and nowhere is this truer than in AI adoption. Successful AI transformation requires more than the right technology; it demands a culture that embraces innovation and practical business sense.

✓ REALITY CHECK

Myth: Traditional processes are more reliable and cost-effective.

Reality: Hidden costs of traditional methods often exceed AI implementation costs.

Impact: Most businesses lose $20K+ annually through inefficient processes.

AI Pro Tip: Start by calculating the true cost of your manual processes.

> *"Agility is key. So you're going to have to have people who understand the technology, but also understand the business needs, what the customers are looking for. And that's a rare combination."*

ERIK BRYNJOLFSSON
DIRECTOR OF STANFORD DIGITAL ECONOMY LAB [2]

This balance of technical understanding and business acumen doesn't happen by accident. It requires thoughtful leadership and a systematic approach to cultural change. Let me show you how one company successfully navigated this cultural transformation.

CASE STUDY: TRAXX FLOORING'S CULTURAL TRANSFORMATION

INITIAL SITUATION:

- Mid-sized flooring company (25 employees, $8.2M revenue)
- Traditional communication approaches
- Siloed departments with isolated workflows
- Overwhelmed by email volume (85 emails daily per employee)
- Resistant to technological change

IMPLEMENTATION APPROACH:

- Started with a "culture health check"
- Created cross-functional collaboration teams
- Implemented basic AI tools for quick wins
- Established clear success metrics

IMPLEMENTATION TIMELINE:

Month 1-2: Foundation

- Conducted culture health check ($5,000 investment)
- Created cross-department teams
- Basic AI email sorting implemented

Month 3-4: Expansion

- Rolled out team collaboration tools
- Started workflow automation
- Initial ROI: 25% reduction in email time

Chapter 2: Rewiring Your Brain for AI Success

Month 5-6: Integration

- Full AI communication suite deployed
- 50% reduction in email volume
- $180,000 annual productivity savings

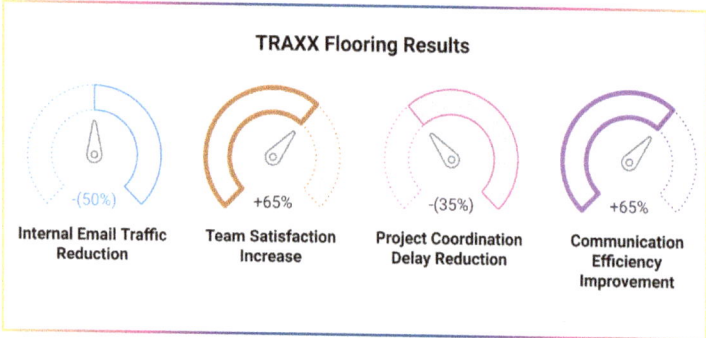

TRAXX Flooring Results

Internal Email Traffic Reduction: -(50%)

Team Satisfaction Increase: +65%

Project Coordination Delay Reduction: -(35%)

Communication Efficiency Improvement: +65%

TRAXX's success illustrates a crucial truth I've observed across dozens of implementations: building an innovation mindset isn't about dramatic changes but consistent, intentional steps in the right direction. Let me share the proven steps that have worked for businesses across industries.

BUILDING AN INNOVATION MINDSET

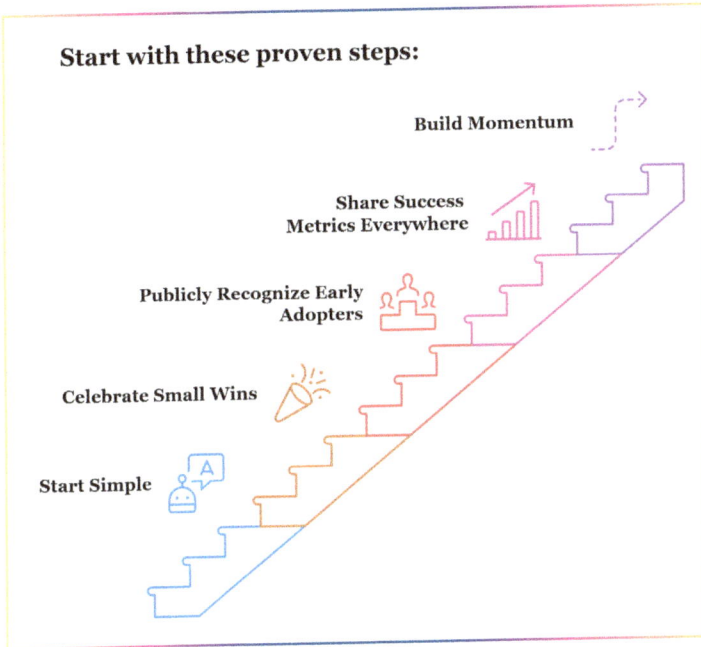

Start with these proven steps:

Build Momentum

Share Success
Metrics Everywhere

Publicly Recognize Early
Adopters

Celebrate Small Wins

Start Simple

Let's examine how successful companies implement each step:

Start Simple

At Green Valley Landscaping, Nina began with just one AI tool for route optimization. *"We wanted to prove the concept before expanding,"* she explained. This focused approach led to a 30% reduction in fuel costs within 90 days.

Celebrate Small Wins

When Duke's Mayonnaise automated their first production line monitoring, they held a team lunch to share the results. *"That first win, just a 12% efficiency gain, created more enthusiasm than any company meeting ever had,"* their operations manager shared. The team began actively looking for new opportunities to implement AI.

Publicly Recognize Early Adopters

TRAXX Flooring created an "AI Champion of the Month" program. Team members who embraced and improved AI processes earned recognition and became mentors to others. *"Our initial skeptics became our biggest advocates,"* Craig noted. Within three months, voluntary participation in AI initiatives rose from 20% to 85%.

Share Success Metrics Everywhere

EndUp Furniture installed digital dashboards showing real-time AI impact on cash flow. *"When everyone could see the daily improvements, it created healthy competition between departments,"* Tom explained. Each team began challenging others to match or exceed their results, driving company-wide adoption.

Build Momentum

Palmetto Credit Union turned its AI journey into a strategic advantage. *"Each success built confidence for the next project,"* Lisa Danforth, VP, shared. Starting with fraud detection, they expanded to customer service, then lending, ultimately transforming their entire operation.

Their systematic approach reduced fraud losses by 75% and improved customer satisfaction by 35%.

These proven steps raise an important question: How ready is your organization to begin this transformation? Through implementing AI across dozens of businesses, I've identified eight critical factors that determine implementation success.

✓ REALITY CHECK

Myth: Innovation requires massive organizational change.

Reality: Successful innovation builds through incremental wins.

Impact: Organizations that rush change face 3x higher failure rates.

AI Pro Tip: Focus on one successful implementation before expanding.

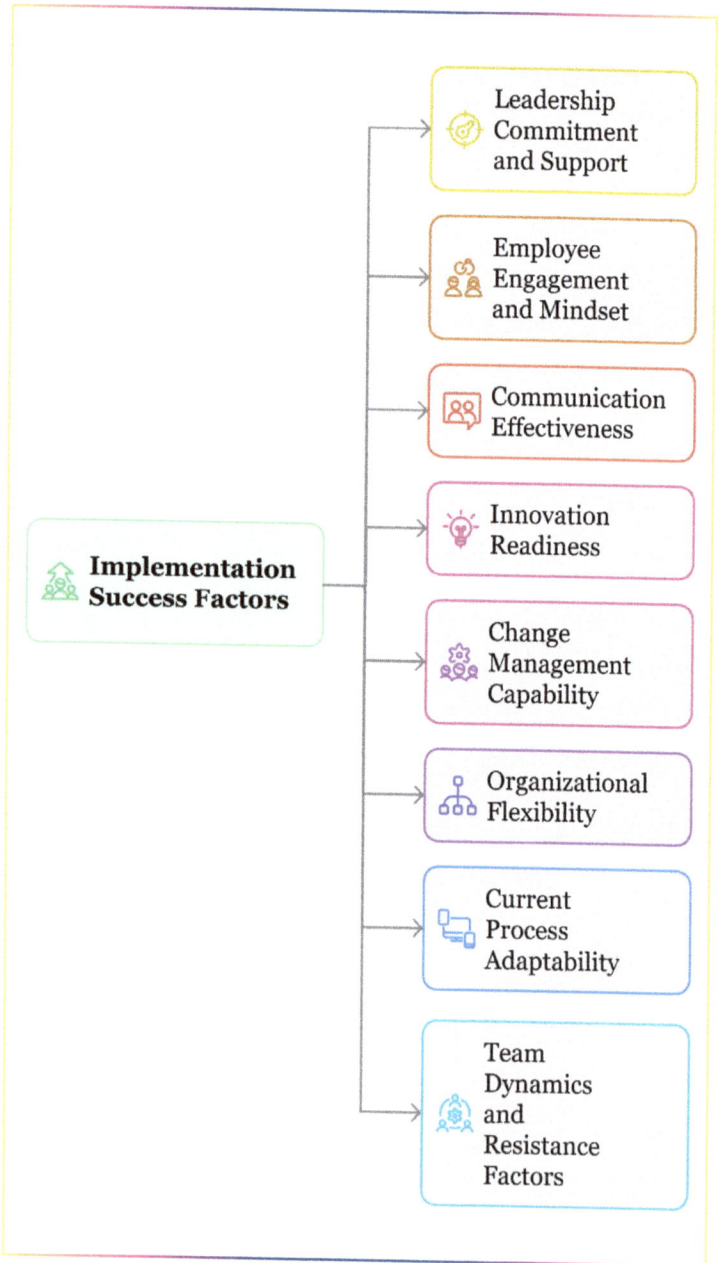

Implementation Success Factors

- Leadership Commitment and Support
- Employee Engagement and Mindset
- Communication Effectiveness
- Innovation Readiness
- Change Management Capability
- Organizational Flexibility
- Current Process Adaptability
- Team Dynamics and Resistance Factors

Real-World Application: Alex N.'s service company achieved remarkable results by:

- Holding monthly innovation meetings
- Creating a simple reward system for improvement ideas
- Starting with basic AI tools and scaling up
- Measuring and celebrating progress

Their results speak for themselves:

- 95% reduction in billing errors
- 40% improvement in employee satisfaction
- 28% increase in customer retention

COMMON PITFALLS TO AVOID

As you begin your AI journey, be mindful of these frequent challenges:

- Trying to do too much too soon
- Failing to prepare your team culturally.
- Focusing on technology without clear business goals
- Ignoring the importance of data quality
- Not measuring the impact of AI initiatives effectively

✓ REALITY CHECK

Myth: Failed AI implementations result from poor technology.

Reality: Most failures stem from inadequate preparation and rushed deployment.

Impact: Organizations that skip preparation face 70% higher failure rates.

AI Pro Tip: Identify and address potential issues before they arise.

LOOKING AHEAD

In our next chapter, I'll break down AI in the simplest terms possible, just as I've done for hundreds of business leaders. No technical jargon, no complicated concepts, just straight talk about what AI really is and how it can work for you.

1. https://elevatesociety.com/quotes-by-brian-herbert/
2. Brynjolfsson, E. (2022). "The Business of Artificial Intelligence." MIT Sloan Management Review, 63(4), 52-59.

CHAPTER 3

WHAT IS AI ANYWAY? THE EASIEST EXPLANATION YOU'LL EVER GET

"Some people call this artificial intelligence, but the reality is this technology will enhance us. So instead of artificial intelligence, I think we'll augment our intelligence."

GINNI ROMETTY
FORMER CHAIRMAN, PRESIDENT, AND CEO OF IBM [1]

✓ REALITY CHECK

Myth: You need to be a technical expert to understand and use AI.

Reality: Basic AI concepts are accessible to any business leader.

Impact: Overcomplicating AI leads to delayed adoption.

AI Pro Tip: Focus on practical applications rather than technical details.

EVERYONE CALLS IT AI, BUT I'M REALLY TALKING ABOUT GEN AI

When my clients talk about 'AI,' they're usually referring to Generative AI, the specialized tool that creates content like text, images, and code. While Gen AI is just a small piece of the AI universe, it's the game-changing technology we'll focus on in this book. I'll use both terms interchangeably as we explore how they can help your business. Here's a visual that shows where Gen AI fits in the bigger AI picture.

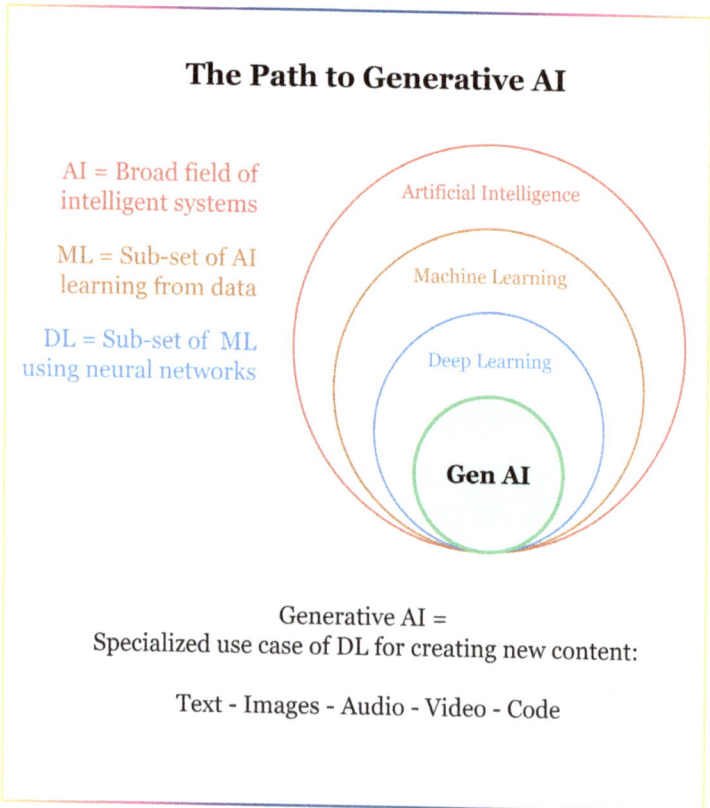

The Path to Generative AI

AI = Broad field of intelligent systems

ML = Sub-set of AI learning from data

DL = Sub-set of ML using neural networks

Artificial Intelligence

Machine Learning

Deep Learning

Gen AI

Generative AI =
Specialized use case of DL for creating new content:

Text - Images - Audio - Video - Code

THE SIMPLEST INTRODUCTION TO
GEN AI...EVER!

Gen AI is a smart tool that learns from lots of data, recognizes patterns, and makes predictions. It gets better over time by learning from its mistakes, just like how people improve through practice.

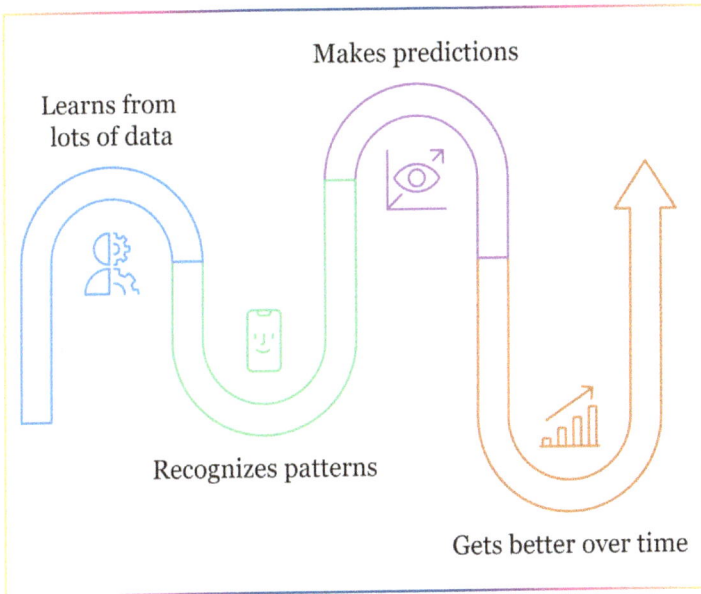

That's Gen AI in a nutshell. Its power for your business lies in its ability to generate human-like content and handle complex tasks 24/7 without getting tired or distracted. To harness Gen AI effectively, let's examine its core capabilities and see what makes it such a valuable asset.

"Generative AI is the most powerful tool for creativity that has ever been created. It has the potential to unleash a new era of human innovation"

ELON MUSK [1]

KEY COMPONENTS OF GEN AI

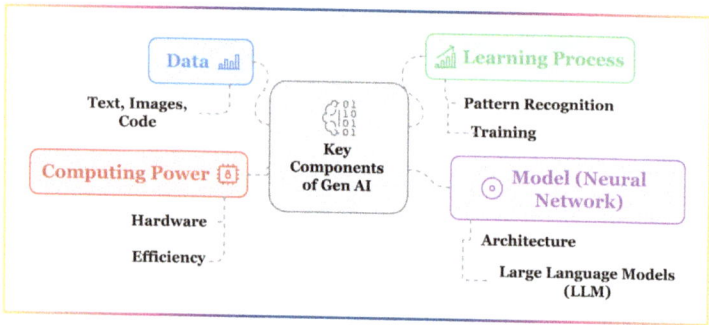

DATA:

Let me explain these components the way I did to Sarah at Moriarty's Books. I told her to think of AI as a new employee. Just like a person, it needs certain things to learn and perform effectively.

First, there's data. Think of this as AI's textbooks. When Sarah showed me her bookstore's data, it included everything from daily sales records to customer preferences. *"But this is just basic business information,"* she said. Exactly! Your AI system learns from the everyday information your business already collects. It takes what your company knows about customer behaviors, operational patterns, and market trends and then turns that information into actionable insights.

LEARNING PROCESS:

AI really shines in the learning process. I explain this to clients using a new employee analogy: Imagine hiring someone who learns from every transaction and remembers every detail perfectly. That's how AI's learning process works.

When Manual G.'s accounting firm implemented its first AI system, they were skeptical about how it would learn their complex processes. *"We handle thousands of transactions with hundreds of variables,"* he told me. But that's precisely what makes AI powerful. The learning process continuously:

- Identifies patterns across those thousands of transactions
- Refines its understanding with each new piece of information
- Adapts to changing conditions automatically
- Applies past lessons to new situations

Think of it as having a team member who gets better every single day without ever forgetting a lesson learned.

MODEL (NEURAL NETWORK):

Think of an AI model (neural network) like a highly experienced business mentor who has learned from thousands of previous situations. Just as a seasoned mentor can:

- Remember every similar situation they've encountered
- Consider multiple factors simultaneously
- Apply past lessons to new challenges
- Make connections between seemingly unrelated experiences

That's what your AI model does. But instead of decades of personal experience, it processes millions of examples at incredible speed.

Like a mentor who becomes better at advising with each consultation, your AI model improves its understanding with each interaction.

COMPUTING POWER:

Finally, let's discuss computing power—the engine that makes everything run. Many business owners worry about this, imagining they'll need a room full of servers. Let me provide a real example to dispel those fears.

When Sarah at Moriarty's Books started her AI journey, she ran the entire system on their existing computers and cloud services. *"But isn't AI incredibly resource-intensive?"* she asked. The truth is that modern AI solutions are remarkably efficient. Think of computing power like electricity for your business:

- You don't need to generate your own; you can plug into existing infrastructure
- You only use (and pay for) what you need
- It scales up or down with your business demands
- Cloud services handle the heavy lifting

"The best part," Sarah told me later, *"was that we didn't have to buy any new equipment. Everything runs through our regular business computers and the cloud."*

✓ REALITY CHECK

Myth: AI requires expensive computers and software.

Reality: Many AI solutions run on standard business hardware or in the cloud.

Impact: SMBs often overestimate technical requirements.

AI Pro Tip: Start with cloud-based AI solutions that scale with your needs.

IGNORE THE HYPE. HERE'S WHAT AI IS NOT...

After implementing AI across many industries, I've noticed that understanding what AI can't do is just as important as knowing what it can do. Let me share the most common misconceptions I encounter:

"Rich, will AI make all our decisions for us?" This is a common question many business owners ask me.

The answer is no. AI is not a magic solution or a fact machine. Think of it more like a completely trained, well-read assistant. It can provide insights based on patterns it's seen, but sometimes it misremembers or misinterprets things, just like any assistant might.

When the CFO of a manufacturing company expressed concern that AI would take control of its operations, I explained that AI isn't autonomous and needs human oversight and direction. *"Think of it as your intelligent assistant,"* I told her. *It can handle routine tasks, but you must review its work and make the final decisions."*

Here are other crucial limitations I make sure my clients understand:

AI ISN'T PERFECT

One of my retail clients learned this early on. Their AI made inventory predictions based on historical data but couldn't account for unexpected events like local festivals. That's why human oversight remains crucial.

AI ISN'T CREATIVE ON ITS OWN

"But I thought AI would generate all our marketing content!" a client once told me. I explained that while AI can help remix existing ideas in new ways, it's more like a DJ mixing existing songs than a musician creating original melodies.

AI ISN'T EMOTIONALLY INTELLIGENT

A customer service manager once asked me if AI could handle sensitive customer complaints. While AI can recognize words associated with emotions and respond with appropriate templates, it doesn't truly understand feelings. It's like having an assistant who can spot an upset customer and follow protocol but cannot empathize with a human.

AI CAN'T TRULY UNDERSTAND CONTEXT

"But will it understand our industry's unique situations?" This question came from a healthcare administrator. I explained that while AI can process industry terminology and follow patterns, it doesn't "get" context as humans do. It's like trying to teach someone a local dialect using only a textbook; they might learn the words but miss the cultural nuances.

AI ISN'T SELF-AWARE

Despite what Hollywood and science fiction suggest, AI has no consciousness or self-awareness. When a tech startup founder worried about AI developing its own agenda, I explained that AI is more like a sophisticated smartphone: It responds to commands and executes tasks but has no inner thoughts or feelings.

AI CAN'T REPLACE HUMAN JUDGMENT

This is perhaps the most important point I emphasize with my clients. A financial services firm I worked with put it best: *"We use AI to analyze market trends and flag potential opportunities, but we still need our analysts to make the final investment decisions."* AI can process vast amounts of data and suggest options but can't replace human wisdom, especially in complex or ethical decisions.

AI ISN'T A SET-AND-FORGET SOLUTION

One of my manufacturing clients learned this the hard way when they started seeing declining results after 90 days. They assumed that once their AI system was installed, it would operate on autopilot. However, AI needs regular monitoring, updating, and refinement. It's more like having a new team member who needs ongoing training and supervision than installing a simple automation tool.

WINS AND WEAKNESSES

Understanding what AI can't do is just part of the picture. Let me share concrete examples of what Gen AI can and cannot do in your business. This comes from my experience helping business owners set realistic expectations for their AI implementations:

CAN DO:

- Write a first draft of your marketing email, but you'll need to review and personalize it
- Generate ideas for social media posts based on your business goals

- Summarize long customer feedback into key themes and actionable points
- Help structure your business proposal following industry standards
- Create basic code templates that your developers can customize

CANNOT DO:

- Create factually perfect content without human verification
- Generate completely original images (it combines elements it's learned)
- Understand your business's unique context without specific guidance
- Make strategic decisions that require real-world judgment
- Guarantee accuracy in rapidly changing or specialized fields

Understanding these limitations leads to a question I often hear from business leaders: *"Rich, how exactly is AI different from human intelligence?"* Let me break down what I've learned:

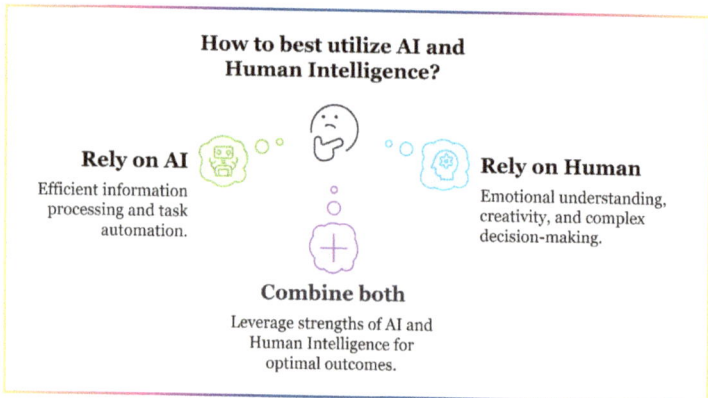

How to best utilize AI and Human Intelligence?

Rely on AI
Efficient information processing and task automation.

Rely on Human
Emotional understanding, creativity, and complex decision-making.

Combine both
Leverage strengths of AI and Human Intelligence for optimal outcomes.

NO, AI ISN'T JUST A SMARTER VERSION OF US

One of the keys to successful AI implementation is understanding that AI doesn't think like we do. Let me show you the fundamental differences I've observed:

LEARNING

When I explain this to clients, I often use the following comparison: While humans learn from a single experience, like touching a hot stove once, AI needs thousands of examples to *"learn"* that heat is dangerous. AI processes massive amounts of data to spot patterns but doesn't truly understand what it's learning. Humans don't just process information; we integrate it with our experiences, emotions, and instincts.

UNDERSTANDING CONTEXT

"But will it understand our unique business situation?" a client recently asked me. The answer is nuanced. While AI can process industry-specific information, it doesn't grasp context the way we do. Think about how you understand when a customer is making a joke rather than being serious; AI still struggles with that.

CREATIVITY

"Can't AI just create all our marketing content?". While AI can generate variations based on existing patterns, it's more like a skilled editor than a true creator. I explain to clients that AI can help brainstorm ideas or optimize content, but those breakthrough creative insights? Those still come from your human team's ability to think outside the box and connect emotionally with customers.

Chapter 3: What is AI Anyway? The Easiest Explanation You'll Ever Get

EMOTIONS AND EMPATHY

A human resources manager once asked me if AI could be used to deliver performance reviews. My answer was simple: AI can compile data, analyze trends, and even suggest constructive feedback, but it can't deliver the nuances of encouragement or understand the impact of tough conversations. It's like having a data analyst explain performance metrics without the warmth or intuition to inspire growth or soften criticism.

PROBLEM-SOLVING

One of my manufacturing clients put it best: *"AI is amazing at optimizing our production schedule, but it completely froze when we had an unexpected supply chain crisis."* Exactly. AI excels at structured problems with clear rules, but human intelligence is irreplaceable in complex, ambiguous situations requiring judgment calls.

FLEXIBILITY

"But I thought AI could do everything!" I hear this a lot. I explain it this way: AI is like a highly specialized professional, excellent within its domain but struggling when asked to step outside it. On the other hand, humans can pivot quickly, applying lessons from one situation to completely different scenarios.

MORAL AND ETHICAL DECISIONS

A healthcare executive once asked me if AI could make treatment priority decisions. This highlights a crucial distinction between AI and human intelligence. While AI can process protocols and procedures, it cannot weigh moral implications. It follows rules but doesn't understand ethics.

CONSCIOUSNESS

When clients worry about AI becoming too powerful, I remind them of this fundamental difference: AI is a sophisticated tool, not a conscious being. It's more like a highly advanced calculator than a thinking entity. It processes information but doesn't reflect on its existence like humans do.

Duke's Mayonnaise made this distinction clear when it implemented its quality control system. While AI could analyze thousands of data points per second to detect potential issues, it took human expertise to understand the implications of its unique recipes and customer preferences. Combining AI analysis and human judgment led to a 13% improvement in quality consistency.

SUMMARY OF DIFFERENCES

In my experience, AI excels at specific, data-driven tasks. It's unbeatable at analyzing patterns in massive datasets or automating repetitive processes. But humans bring irreplaceable qualities like intuition, emotional intelligence, and ethical judgment to the table. The goal isn't to make AI think like humans but to leverage each of its unique strengths.

Speaking of strengths, this brings us to what really powers AI's capabilities: Data.

Remember how I mentioned AI needs thousands of examples to learn? Let me show you why the quality and quantity of your data can make or break your AI implementation.

THE ROLE OF DATA

Think of data like ingredients in a recipe. Just as a chef needs quality ingredients, AI needs quality data. Rachel G., a restaurant manager I worked with, transformed her scheduling efficiency using just three months of clean, well-organized data. *"We didn't need years of history,"* she told me. *"We just needed the right data, organized the right way."* The key point? Quality trumps quantity.

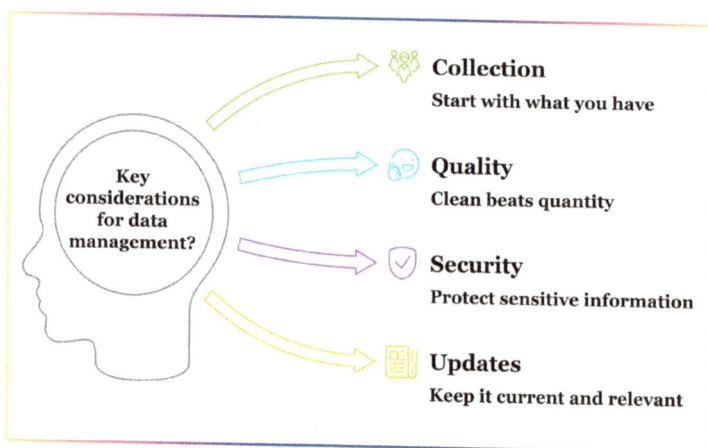

When explaining AI technologies to business owners, I use the "toolbox approach." Just as you wouldn't use a hammer for every home repair, different AI tools serve various business purposes. Let me show you what's in the toolbox:

THE AI TOOLBOX

Machine Learning (ML): Consider ML your business's learning system. It's like having an analyst who:

- Never sleeps
- Learns from every transaction
- Spots patterns humans might miss
- Gets smarter over time

Natural Language Processing (NLP): This is how AI understands and responds to human language. It's the technology behind:

- Customer service chatbots
- Email response systems
- Voice assistants
- Document analysis

Computer Vision: This is AI's way of "seeing" and understanding images and videos. It's already helping small businesses with:

- Quality control in manufacturing
- Security monitoring
- Inventory tracking
- Customer behavior analysis

Robotic Process Automation (RPA): This software lets businesses create and deploy "bots" to mimic human actions in digital systems. This is where Manuel G.'s accounting firm started its AI journey. RPA handles repetitive tasks like:

- Data entry
- Report generation
- Invoice processing
- Compliance checking

Myth: RPA will replace your administrative staff.

Reality: RPA frees up staff to focus on higher-value work.

Impact: Staff often resist RPA out of job security concerns.

AI Pro Tip: Focus on how RPA enhances rather than replaces human work.

WANT TO EXPLORE MORE AI TERMS?

I've found that keeping things simple and focused is crucial for success with AI. That's why I've intentionally limited our AI toolbox to just these four essential technologies. These workhorses drive real business value - no fancy jargon or bleeding-edge concepts that look impressive but deliver little practical benefit.

Tool#1 - AI Terms Glossary

However, some of you might want to explore AI terminology more deeply. For those curious about other AI concepts and terms you might encounter on your journey, I've created a comprehensive online reference guide that breaks down AI terms into clear, business-focused explanations. You'll find real-world examples showing how each concept applies to actual business situations and practical use cases from successful implementations. This guide will help you cut through the technical jargon and understand exactly what different AI terms mean for your business.

You can explore this reference guide at your own pace.

THE AI TRANSFORMATION ACROSS BUSINESS FUNCTIONS

Let's look at how these technologies in the AI Toolbox are changing specific business areas:

Sales and Marketing

Lead Prediction

Message Personalization

Customer Behavior Analysis

Real-Time Pricing Optimization

-AI forecasts which sales leads are most likely to convert into customers.

-AI tailors marketing messages to individual customer preferences.

-AI examines and interprets customer behavior data to identify trends.

-AI adjusts pricing strategies dynamically based on market conditions.

Operations and Supply Chain

Inventory
Optimization

Logistics
Efficiency

Predictive
Maintenance

Waste
Reduction

-Ensures optimal stock levels to meet demand without excess.

-Enhances the movement of goods to reduce delays and costs.

-Anticipates and addresses maintenance needs to prevent downtime.

-Minimizes waste and associated costs through efficient practices.

Human Resources

Candidate
Matching

Employee
Engagement

Recruitment
Streamlining

Task
Automation

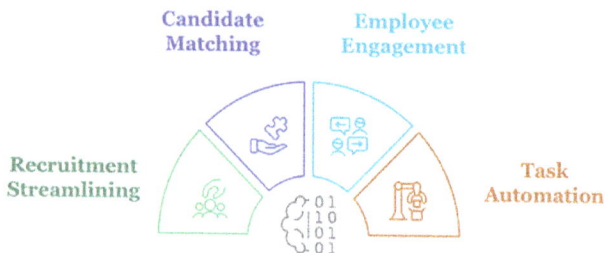

-Ensures the best fit between candidates and job roles.

-Boosts morale and productivity through active involvement.

-Optimizes the hiring process to attract top talent efficiently.

-Reduces manual workload by automating repetitive tasks.

Finance and Accounting

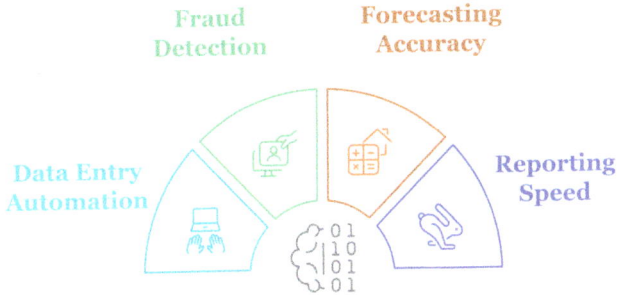

Fraud Detection

Forecasting Accuracy

Data Entry Automation

Reporting Speed

-Identifies and alerts on suspicious financial activities.

-Enhances predictive models for better financial planning.

-Streamlines processes by reducing manual input errors and time.

-Accelerates the generation and distribution of financial reports.

✓ REALITY CHECK

Myth: AI implementation requires a complete business overhaul.

Reality: AI can solve specific business problems while integrating with existing operations.

Impact: Starting with focused problems leads to quick wins.

AI Pro Tip: Begin with a single pain point that directly impacts your bottom line.

CASE STUDY: MORIARTY'S BOOKS AI JOURNEY

When Sarah and I first discussed AI for her bookstore, she wasn't looking to revolutionize her business. She just wanted to solve a persistent inventory problem. *"Rich, I'm tired of telling customers we don't have books they want while our storage room is full of books nobody's buying."*

INITIAL SITUATION:

- Manual inventory tracking
- 35% stock-out rate on popular titles
- Excess inventory tying up capital
- Limited customer purchase pattern insights
- Staff overwhelmed with manual processes

IMPLEMENTATION JOURNEY:

Data Assessment Phase

- Collected three years of sales data
- Organized customer purchase history
- Cataloged current inventory
- Mapped seasonal trends

Solution Design

- Started with basic AI inventory management
- Added customer behavior analysis
- Integrated point-of-sale data
- Implemented predictive ordering

Staff Training and Adoption

- Hands-on system training

- Regular feedback sessions
- Performance monitoring
- Continuous improvement cycles

RESULTS:

- 30% reduction in excess inventory
- Significant decrease in stock-outs
- 15% increase in overall sales
- 20% improvement in cash flow
- Staff time was redirected to customer service

"The key was starting small but thinking big," Sarah told me. *"We didn't try to solve everything at once, but each improvement led naturally to the next."*

✓ REALITY CHECK

Myth: You need complete AI knowledge before starting.

Reality: Understanding grows through implementation.

Impact: Waiting for perfect knowledge delays benefits.

AI Pro Tip: Start with basics; learn by doing.

COMMON PITFALLS TO AVOID

- Overestimating AI capabilities and expecting unrealistic results → Understand what AI is not.
- Setting unrealistic timelines → Start with 90-day pilot projects
- Choosing the wrong AI tools → Begin with proven, industry-specific solutions
- Poor data preparation → Start data cleaning 30 days before implementation
- Inadequate training → Plan for 3-5 hours per employee initially

Chapter 3: What is AI Anyway? The Easiest Explanation You'll Ever Get

OPTIONAL STEPS:

1. Browse our comprehensive <u>Tool #1</u> AI Terms Glossary.
2. Revel in that you now understand basic AI without all the "hype and fluff"!

LOOKING AHEAD

In our next chapter, we'll explore how to navigate the critical change management aspects of AI implementation. As you'll see, understanding the technology is just the first step. The real success comes from how you implement it in your organization.

Remember: Your goal isn't to become an AI expert. Your goal is to understand enough to make informed decisions about how AI can help your business grow. The frameworks and examples I've covered give you that foundation.

Understanding AI is important, but the hardest part of successful implementation isn't the technology itself. In my experience, people, not technology, determine success or failure.

In our next chapter, I'll show you exactly how successful businesses manage this human element. Their experiences will help you navigate one of the most crucial aspects of AI adoption: leading your team through change.

[1] https://www.forbes.com/sites/nicolemartin1/2019/06/27/13-greatest-quotes-about-the-future-of-artificial-intelligence/

CHAPTER 4

THE HARDEST PART OF AI ISN'T AI...IT'S US

*"The biggest challenge of AI adoption isn't the technology, **it's the people**. Leaders must focus on creating a culture that embraces change, fosters continuous learning, and sees AI as a tool to augment human capabilities rather than replace them. Only then can organizations fully realize the transformative potential of AI."*

CHARLENE LI, FOUNDER AND SENIOR FELLOW AT ALTIMETER [1]

When Nina B. from Green Valley Landscaping and I initially discussed implementing AI, her voice was filled with excitement and apprehension. *"Rich, I know AI can help us grow, but how do I get my operations team on board? They're worried it'll make their jobs obsolete."*

I've heard this concern countless times, and here's what I always say: Implementing AI isn't just about new technology; it's about transforming how your entire organization operates. And that transformation starts with your people.

Let me explain how we transformed Nina's 200+ employees from skeptics to enthusiasts and how you can do the same in your business.

✓ REALITY CHECK

Myth: AI implementation is primarily a technical challenge.

Reality: The biggest hurdles in AI adoption are often cultural and organizational.

Impact: Focusing solely on technology leads to poor adoption and wasted resources.

AI Pro Tip: Prioritize change management from the start of your AI journey.

GETTING YOUR BUSINESS AI-READY THE RIGHT WAY

Before we even touched a single AI tool at Green Valley, we assessed their readiness for change. Here's what I've learned works best:

1. **Cultural Readiness**: Is your team open to new ideas? At Green Valley, we found a mix. Some employees were excited about AI, and others were wary. That's normal. Your

goal isn't 100% enthusiasm from day one; it's creating an environment where people feel safe asking questions and expressing concerns.

2. **Technical Readiness**: Can your current systems support AI? Nina was surprised when I asked about their existing software. *"But I thought we'd need all new systems for AI,"* she said. Not necessarily. Often, we can integrate AI with your current setup, saving time and money.

3. **Skills Readiness**: Does your team have the right skills or the ability to learn them? This isn't about turning your office team into data scientists but assessing adaptability. At Green Valley, we found that many employees were eager to learn new skills once they understood how they would make their jobs easier.

4. **Data Readiness**: Do you have the right data available? *"But we're just a landscaping company,"* Nina protested. *"We don't have fancy data."* You'd be surprised. Every customer interaction, every job completed, and every route driven is valuable data. We just need to organize it effectively.

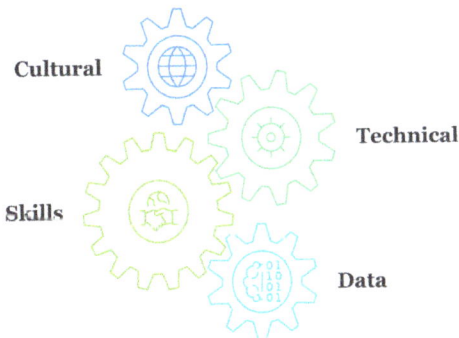

Types of AI Readiness for Effective Change

Cultural

Technical

Skills

Data

✓ **REALITY CHECK**

Myth: You need perfect readiness before starting with AI.

Reality: Readiness is a journey, not a destination.

Impact: Waiting for perfect conditions often leads to missed opportunities.

AI Pro Tip: Start where you are and grow your readiness alongside your AI implementation.

CRAFTING A VISION THAT MAKES AI EXCITING FOR EVERYONE

Here's where many business leaders stumble. They get excited about AI's potential but struggle to communicate that excitement to their team. Let me show you how we crafted a vision at Green Valley that got everyone on board:

Align with Your Business Strategy

Nina wanted to grow the business without sacrificing quality, so we framed AI as a tool to "serve more clients with the same Green Valley touch."

Address Specific Pain Points

We highlighted how AI could reduce the routing team's planning time and make the maintenance crews' schedules more efficient.

Make It Easy to Understand

We avoided technical jargon. Our vision statement was simple: "Use AI to help every Green Valley employee work smarter, not harder, so we can grow our business while improving our service."

Inspire and Motivate

We illustrated how AI could make everyone's job more interesting by handling routine tasks and freeing up time for more engaging work.

The result? A vision that resonated across the company: "Leveraging AI to provide personalized landscaping experiences for every client, improving satisfaction and loyalty while increasing operational efficiency."

✓ REALITY CHECK

Myth: Employees will naturally see the need for AI.

Reality: Many will be comfortable with the status quo.

Impact: Lack of urgency leads to half-hearted adoption.

AI Pro Tip: Clearly communicate why AI is crucial for the company's future.

But I've Been Doing It This Way for 8 Years!

Let me share a critical lesson about the human side of AI implementation. Plush Leg Warmers reached out after their first attempt at AI pricing optimization ended in an open rebellion. Their head of pricing, Katie, had built their entire pricing strategy over eight years, priding herself on her intuitive understanding of their market. When management invested $35,000 in an AI pricing system, Katie saw it as a direct challenge to her expertise and thought it was the company's way of "showing her the door."

Instead of embracing the new tools, Katie spent weeks documenting every perceived mistake in the AI's pricing suggestions. She rallied other team members to reject the system, arguing that *'algorithms can't understand our customers like we do.'* The AI system, which could have been analyzing thousands of data points to optimize their pricing, sat unused while Katie continued making manual adjustments based on gut feel.

When they brought me in, they had a demoralized team, wasted investment, and growing friction between management and staff. The solution wasn't better technology; it was better change management. We worked with Katie to make her the project champion, combining her market insights with AI's analytical power. Within two months, their profit margins increased by 7% without losing customers. Katie now leads training sessions on AI-enhanced pricing strategy.

'I went from seeing AI as my replacement to seeing it as my superpower,' Katie later admitted.

Developing a Change Management Strategy

Stories like Katie's at Plush Leg Warmers taught me that even the most powerful AI tools fail without proper change management. Through dozens of implementations, I've found that combining two classic change models creates a road map that turns skeptics into champions. Let me show you how these models work together to guide successful AI integration:

KOTTER'S 8-STEP CHANGE MODEL IN ACTION

1. **Create Urgency**—I helped Nina present AI not as a threat but as an opportunity. *"If we don't adapt,"* she told her team, *"our competitors will leave us in the dust."* This wasn't fear-mongering; it was reality.

2. **Form a Powerful Coalition**—We didn't just rely on top-down mandates. We identified influencers at all levels of Green Valley, from veteran landscapers to tech-savvy office staff. This diverse group became our AI ambassadors.

3. **Create a Vision for Change**—Remember that compelling vision we crafted? This is where it came into play, giving everyone a clear picture of our direction.

4. **Communicate the Vision**—Nina regularly discussed AI in team meetings, company newsletters, and casual conversations. *"I felt like a broken record,"* she admitted, *"but it worked."*

5. **Remove Obstacles**—We identified and addressed barriers to AI adoption. For instance, when we realized some team members lacked basic computer skills, we set up training sessions.

6. **Create Short-Term Wins**—This is crucial. We started with a small AI project, optimizing routes for two maintenance crews. It created a buzz across the company when they reported saving an hour and a half a day, having more time for lunch, and having less drive time.

7. **Build on the Change**—Using that momentum, we expanded AI into scheduling and then customer service. Each success built excitement for the next phase.

8. **Anchor the Changes in Corporate Culture**—Eventually, using AI became *"just how we do things"* at Green Valley. It was no longer new tech but an integral part of their operations.

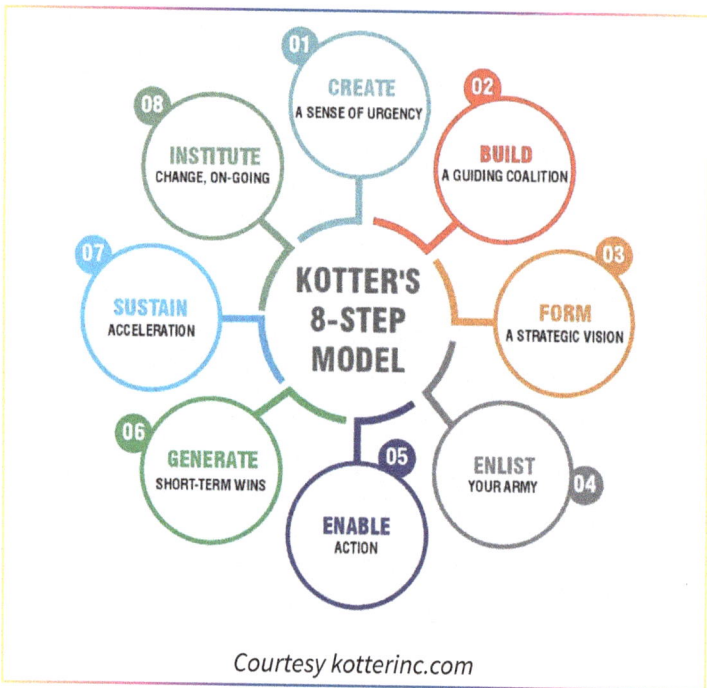

Courtesy kotterinc.com

LEWIN'S CHANGE MANAGEMENT MODEL: UNFREEZE, CHANGE, REFREEZE

While Kotter's model gave us a detailed road map, Lewin's simpler model helped us understand the emotional journey:

1. **Unfreeze**—We had to thaw out old habits and mindsets. This meant acknowledging that change could be uncomfortable but necessary.

2. **Change** —We rolled out new AI tools and processes in the active implementation phase.

3. **Refreeze**—Once changes were in place, we worked to solidify them as the new normal.

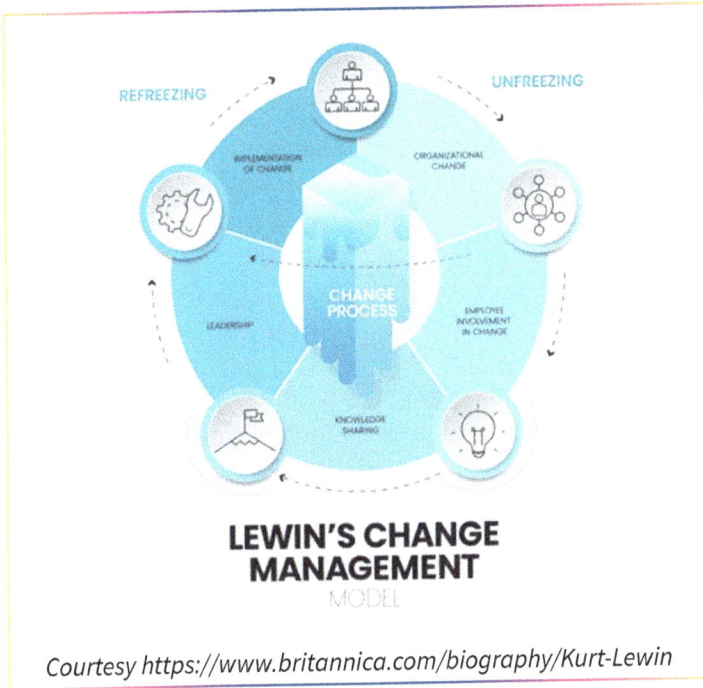

LEWIN'S CHANGE MANAGEMENT
MODEL

Courtesy https://www.britannica.com/biography/Kurt-Lewin

"The Lewin model helped me understand why some team members struggled at different points," Nina told me later. *"It wasn't resistance— it was a natural part of the change process."*

Supporting Employees Through the Transition

Now, let's talk about the heart of change management: your people.

At Green Valley, we focused on four key areas:

1. **Addressing Fears and Concerns**—We created safe spaces for employees to voice their worries. The most common was fear of job loss. We countered this by showing how AI would enhance their roles, not replace them.

2. **Providing Training and Development**—We set up a comprehensive training program, from 'AI 101' sessions for everyone to specialized training for key roles.

3. **Creating Feedback Channels**—We established multiple ways for employees to share their thoughts and experiences, from anonymous suggestion boxes to regular check-ins.

4. **Celebrating Early Wins**—Remember that route optimization success? We made sure everyone knew about it, showing the tangible benefits of AI adoption.

✓ REALITY CHECK

Myth: Employees will resist AI no matter what.

Reality: Most employees embrace AI when they understand how it benefits them.

Impact: Assuming resistance can become a self-fulfilling prophecy.

AI Pro Tip: Focus on education, involvement, and demonstrating clear benefits.

In our next section, I'll share how we built a coalition of AI champions at Green Valley and how you can do the same in your business.

Successful AI integration isn't about technology; it's about people. And nobody knows your people better than you do.

Building a Coalition of AI Champions

"Rich, I can't be everywhere at once," Nina told me about a month into Green Valley's AI journey. *"How do I keep the momentum going?"*

This is where building a coalition of AI champions becomes crucial.

Here's how we did it at Green Valley and how you can do it in your business:

Identify Potential Champions

First, we looked for employees who showed natural enthusiasm for AI projects. But here's a tip: don't just focus on the tech-savvy. Some of our best champions were longtime employees who saw AI's potential to solve long-standing problems.

For instance, Jake, a veteran foreman, became an unexpected advocate when he realized AI could optimize watering schedules. *"I've been saying for years we need a smarter way to do this,"* he told me. *"I didn't know AI was the answer, but I'm all in now."*

Empower Your Champions

Identifying champions is just the start. You need to give them the tools and authority to drive change. Here's what worked at Green Valley:

- Extra Training: We provided our champions with advanced AI training, not to make them experts but to give them confidence in discussing AI with their peers.
- Resources: First, we gave champions access to AI tools, allowing them to experiment and showcase their benefits to their colleagues.
- Authority: We empowered champions to decide how to implement AI in their areas. This ownership drove engagement.

Create Opportunities for Showcasing Success

We set up regular "AI in Action" sessions where champions could demonstrate their AI wins to the rest of the company. This motivated the champions and inspired others to get involved.

Lisa, the HR manager at a tech startup I worked with, used a similar approach. *"We created an 'AI Innovation Award',"* she told me. *"It became a friendly competition to see who could find the most impactful AI application in their role."*

Recognize and Reward Efforts

Recognition doesn't always have to be formal or financial. At Green Valley, we highlighted the efforts of our AI champions in company meetings and newsletters. This public acknowledgment motivated continued engagement.

✓ REALITY CHECK

Myth: Only tech-oriented employees can be AI champions.

Reality: Enthusiasm and problem-solving skills are more important than technical knowledge.

Impact: Overlooking non-technical champions can slow adoption across the organization.

AI Pro Tip: Look for problem solvers and influencers in all departments.

Addressing Ethical Concerns Related to AI

New concerns emerged as we rolled out more AI initiatives at Green Valley. *"Rich, some of our clients are asking about data privacy,"* Nina told me. *"And I overheard some employees worrying about AI making unfair employee scheduling decisions."*

These are valid concerns, and addressing them head-on is crucial to building trust in your AI initiatives. Here's how we tackled it:

Develop Clear Ethical Guidelines

We created a straightforward AI ethics policy that covered the following:

- Data privacy and security

- Fairness and bias prevention
- Transparency in AI decision-making
- Human oversight and intervention protocols

A copy of this template is available HERE.

Tool #2 - The AI Use Policy Template

Every organization needs clear guidelines for AI use. The AI Use Policy Template provides a framework that covers everything from tool selection to data handling and security protocols. Instead of spending thousands on legal fees or risking compliance issues, you can customize this template to create comprehensive policies that protect your organization.

COMMUNICATE TRANSPARENTLY

We didn't just create these guidelines and file them away. We made them a central part of our AI communications. In every AI-related meeting, we reinforced our commitment to ethical AI use.

Provide Ethics Training

We incorporated AI ethics into our training programs, ensuring everyone understood not just how to use AI but how to use it responsibly.

Establish an Ethics Review Process

We set up a simple review process for major AI initiatives to ensure they aligned with our ethical guidelines. This wasn't about creating bureaucracy but about building trust.

The Results

Within a year of starting this change management journey, Green Valley saw remarkable results:

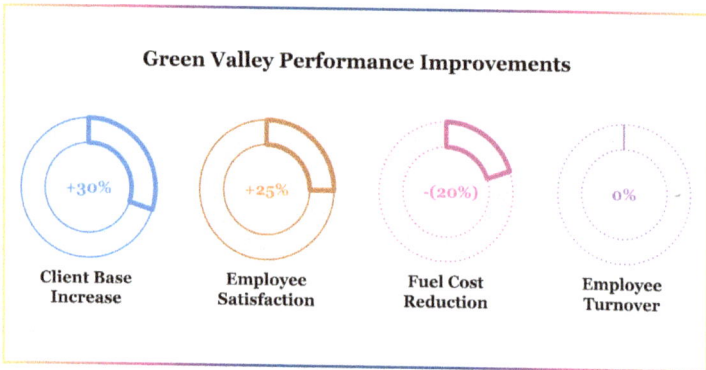

Green Valley Performance Improvements

Client Base Increase	Employee Satisfaction	Fuel Cost Reduction	Employee Turnover
+30%	+25%	-(20%)	0%

But the most telling result? When I asked Nina about the next steps, she said, "*Rich, my team is coming to me with ideas for new AI projects. They're driving this now.*"

And that's the goal of effective change management: to reach a point where AI adoption becomes self-sustaining, driven by your team's enthusiasm and innovation.

Remember, integrating AI into your business isn't just about the technology. It's about people, culture, and, most importantly, how you lead your team through this transformation.

Your AI journey is unique to your business, but the principles discussed here can guide you toward success. Remember, my team and I are always here to help you navigate these changes.

✓ **REALITY CHECK**

Myth: Change management plans need to be complex and comprehensive.

Reality: Start with a simple, clear plan and refine as you go.

Impact: Over-planning can delay important first steps.

AI Pro Tip: Begin with essential elements and build from there.

COMMON PITFALLS TO AVOID

- Underestimating the importance of change management in AI adoption
- Failing to address employee concerns and fears adequately
- Neglecting to provide sufficient training and support
- Communicating inconsistently or infrequently about the AI initiative
- Trying to change too much too quickly
- Not celebrating early wins and milestones
- Failing to align AI initiatives with overall business strategy

Remember: Start small, but think big. You don't need to implement everything at once. Choose the most relevant actions for your business and begin there.

Chapter Summary: When helping SMBs implement AI, I've learned that success depends more on how you manage change than on the technology itself. Let's recap the key lessons from this chapter:

KEY TAKEAWAYS:

START WITH YOUR PEOPLE

- Green Valley's success came from engaging employees early
- Build a coalition of champions across departments
- Address fears and concerns openly and honestly
- Celebrate early wins to build momentum

FOLLOW A STRUCTURED APPROACH

- Use Kotter's eight steps as your road map
- Apply Lewin's model to understand the emotional journey
- Create clear communication channels
- Measure and track progress

BUILD TRUST THROUGH ETHICS

- Develop clear AI guidelines
- Maintain transparency in implementation
- Ensure data privacy and security
- Keep humans in the loop

Remember Nina's words: *"Once we got the people part right, the technology part became much easier."*

✓ REALITY CHECK

Myth: Technology is the hardest part of AI implementation.

Reality: People and processes determine success or failure.

Impact: Focusing solely on technology leads to poor adoption.

AI Pro Tip: Invest equal time in change management and technical implementation.

LOOKING AHEAD

In the next chapter, we'll discuss specific types of resistance you might encounter and how to overcome them. But remember, the foundation we've covered here—engaging your team, building trust, and managing change effectively—will be crucial for every step of your AI journey.

Even with the best change management approach, you'll still face skepticism and resistance. This isn't a sign of failure. It's a natural part of a transformation that every successful business I've worked with has faced and overcome.

In our next chapter, I'll share specific strategies that turned some of my most skeptical clients into enthusiastic AI advocates. Their stories will show you exactly how to address concerns, build confidence, and create momentum for your AI implementation.

[1] https://hrexecutive.com/on-its-own-ai-is-just-tech-this-expert-says-true-transformation-is-the-people-part/

CHAPTER 5

WINNING OVER THE SKEPTICS (YES, EVEN YOU)

"Many people have a failure of imagination and assume we'll use AI to produce the same things but with fewer workers. In fact, if you look through history, most technologies have ended up complementing humans rather than substituting for them." [1]

— ERIK BRYNJOLFSSON, DIRECTOR OF THE STANFORD DIGITAL ECONOMY LAB

Sophia F., the compliance officer at Southwest Healthcare, sat across from me, clearly frustrated. *"Rich, my team thinks AI means they'll be out of a job. How do I convince them that's not true?"*

I've had this conversation dozens of times. In fact, fear of job displacement tops the list of resistance I encountered while implementing AI across a wide range of companies. When handled correctly, AI typically creates more opportunities than it eliminates.

Let me share how we transformed this fear into enthusiasm at Southwest Healthcare and the lessons you can apply to your business.

✓ REALITY CHECK

Myth: AI will replace human workers.

Reality: AI augments human capabilities, often creating new roles and opportunities.

Impact: Fear of job loss creates unnecessary resistance and delays valuable implementation.

AI Pro Tip: Demonstrate how AI enhances rather than replaces human work.

THE THREE FEARS BEHIND AI RESISTANCE

When I meet with leadership teams, I often hear the same concern: *"We've invested years building our expertise. How can AI possibly understand the complexities of our industry?"* This fundamental worry applies to all sectors, whether manufacturing, healthcare, finance, or retail.

This reaction isn't unusual. I've seen similar concerns across dozens of implementations. One Operations Director initially worried about AI replacing her team's decision-making abilities. Yet after implementation, her department achieved 34% faster workflow completion while expanding its strategic capabilities.

The resistance typically stems from three core fears:

JOB SECURITY

"Will this AI make my job obsolete?" This question is asked in nearly every implementation. Yet teams consistently find that AI handles routine tasks, freeing them to focus on more meaningful work. One retail manager's team reduced administrative tasks by 50%, allowing them to double their customer interaction time.

VALUE AND EXPERTISE

Many professionals worry their years of experience will become irrelevant. A financial executive I worked with had this exact concern. Instead, she discovered AI helped showcase her expertise by handling data analysis, giving her more time for strategic planning. Her team's impact on business growth doubled within six months.

CONTROL AND AUTONOMY

Teams often fear losing control over critical decisions. A risk management director expressed this concern before implementation. After integrating AI, her team's fraud detection improved by 75% while enhancing their analytical capabilities. *"AI didn't take control,"* she noted, *"it gave us better control."*

Fears that Cause Resistance

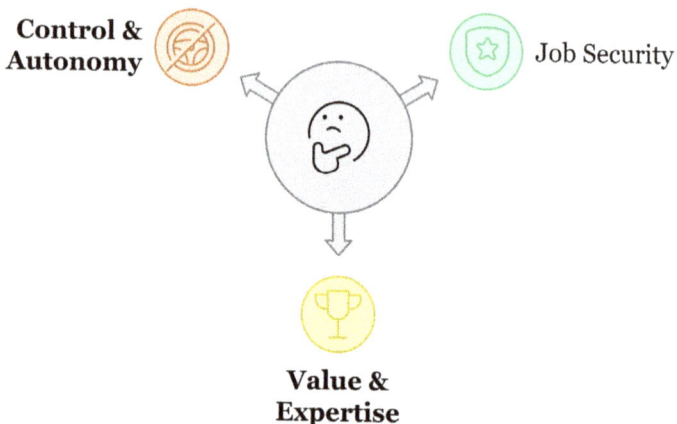

Control & Autonomy

Job Security

Value & Expertise

✓ REALITY CHECK

Myth: AI will replace human workers.

Reality: AI augments human capabilities, often creating new roles and opportunities.

Impact: Fear of job loss creates unnecessary resistance and delays valuable implementation.

AI Pro Tip: Demonstrate how AI enhances rather than replaces human work.

THREE CRITICAL BARRIERS THAT BLOCK AI SUCCESS

Corner Office Cold Feet

When Craig at TRAXX Flooring first proposed implementing AI, his executive team responded lukewarmly. "Our leadership saw AI as a cost, not an investment," he shared. After quantifying potential productivity gains and starting with a small pilot project that showed clear ROI within 45 days, the tone shifted dramatically. Now, his executives ask him to identify new AI opportunities.

Square Peg, Round Hole

Nina at Green Valley Landscaping initially tried implementing an AI scheduling system designed for retail. "We wasted $12,000 on a solution that didn't understand landscaping workflows," she admitted. Learning from this, she focused on solutions built for field services, reducing fuel costs by 30% and improving route efficiency within 90 days.

The "We've Always Done It This Way" Wall

Duke's Mayonnaise faced deep-rooted resistance to changing decades-old production processes. "Our team had legitimate pride in their work," Mike explained. The breakthrough came when they involved veteran employees in the AI implementation process, letting them guide how the technology could enhance rather than replace their expertise. Within six months, these same employees were suggesting new ways to use AI.

In my AI advisory roles, I've learned that recognizing these barriers is just the first step. The real value comes from a systematic approach to address them. Let me show you how Southwest Healthcare transformed these barriers into bridges and the lessons you can apply to your business.

THE SOUTHWEST HEALTHCARE STORY

Let me show you how we turned this resistance into enthusiasm at Southwest Healthcare. Their compliance team was overwhelmed with regulatory monitoring and documentation when we started. They spent countless hours manually reviewing regulations, updating policies, and ensuring compliance.

Their initial reaction to AI implementation? Pure skepticism. But we took a systematic approach to addressing their concerns:

First, we listened. Really listened. I held one-on-one sessions with team members to understand their specific fears and challenges.

Then, we involved them in the solution. Instead of imposing AI tools, we asked them to help design how these tools would integrate into their workflow.

Finally, we started small with a pilot program demonstrating AI's ability to augment rather than replace their expertise.

BUILDING BUY-IN ACROSS ORGANIZATIONS

When Sophia called me three weeks into her healthcare organization's AI implementation, her tone had completely changed. *"Rich, remember how resistant my team was? Well, you won't believe what happened in yesterday's meeting."*

She went on to tell me how one of her most skeptical team members had become an enthusiastic advocate after seeing how AI helped identify a critical regulatory update they initially missed. This transformation pattern repeats across industries, from manufacturing to retail to professional services.

Let me show you the specific strategies that drive successful adoption:

STARTING SMALL WITH CLEAR WINS

Instead of introducing AI across all processes simultaneously, we identify one pain point everyone agrees on. For Sophia's team, it was time-consuming regulatory monitoring. In manufacturing, it might be quality control. For retail, it's often inventory management.

This targeted approach accomplishes two things:

- Demonstrates immediate value
- Allows teams to maintain control over critical decisions

"The key was showing them how AI made them better at their jobs, not replaceable," Sophia told me. Within weeks, her team saw routine monitoring time drop by 75%, freeing them to focus on applying their expertise to complex regulatory interpretations. Similar wins appear across sectors - from retail teams reducing inventory checks to financial analysts spending more time on strategy.

BUILDING TRUST THROUGH INVOLVEMENT

Here's a crucial lesson I've learned: people support what they help create. Our approach includes:

- Creating cross-functional advisory boards
- Incorporating team feedback into AI configuration
- Letting teams decide which processes to automate next

This mirrors what Nina achieved in her operations department. *"Once my team realized they were designing their future, not having it imposed on them, everything changed,"* she shared.

TRAINING AS EMPOWERMENT

One of our most successful strategies is reframing training as career enhancement. Instead of basic *"how to use AI"* sessions, we develop:

- Advanced analysis workshops specific to each industry
- AI-enhanced decision-making courses
- Future career planning that incorporates AI skills

This approach directly addresses the fear of obsolescence by showing how AI makes team members more valuable, not less necessary.

DEMONSTRATING ROI BEYOND COST SAVINGS

While the 30% reduction in compliance costs got leadership's attention, what really won over teams was seeing how AI enhanced their professional capabilities:

- More time for complex analysis
- Fewer errors in routine tasks
- Earlier detection of potential issues
- Enhanced reporting capabilities

As one manufacturing manager noted, *"The real ROI wasn't just in cost savings - it was in how much more effectively we could serve our customers."*

CREATING CHAMPIONS AND ADDRESSING SPECIFIC CONCERNS

Sometimes, the best solutions come from unexpected places. At Southwest Healthcare, our strongest AI champion was Mary, a veteran compliance officer who had initially been one of our biggest skeptics.

"You know what changed my mind, Rich?" she told me during a follow-up visit. *"When I realized AI could handle the mundane monitoring tasks, I finally had time to mentor junior staff, something I'd always wanted to do but never had time for."*

IDENTIFYING AND EMPOWERING CHAMPIONS

Through implementing AI across hundreds of teams, I've learned that effective champions often share these key characteristics:

Experience and Credibility:

- Deep understanding of current processes
- Respected by peers and leadership
- Track record of successful change adoption
- Natural problem-solving abilities

Communication Skills:

- Ability to translate technical concepts
- Strong informal influence with colleagues
- History of effective team collaboration
- Skill at addressing concerns constructively

Leadership Qualities:

- Takes initiative in learning new approaches
- Balances enthusiasm with practicality
- Maintains a positive attitude during challenges
- Helps others overcome obstacles

Growth Mindset:

- Openness to new ways of working
- Curiosity about improvement opportunities
- Willingness to experiment and learn
- Ability to see potential in change

"The best champions aren't necessarily the most technical people," Sophia explained. *"They're the ones who can see the bigger picture and help others see it too."*

ADDRESSING SPECIFIC CONCERNS HEAD-ON

Let me share how we tackled the four biggest concerns at Southwest Healthcare, using approaches I've refined from working with clients:

JOB SECURITY

We created clear career development paths that show how AI will create new opportunities. For example, compliance officers can now focus on strategic risk management rather than routine monitoring.

DATA PRIVACY

This was crucial in healthcare. We:

- Developed transparent data handling protocols
- Created clear accountability structures
- Established human oversight checkpoints
- Implemented robust security measures

DECISION AUTHORITY

We established clear guidelines about which decisions would remain purely human-driven versus AI-assisted. This helped alleviate fears about losing control over critical compliance determinations.

SKILL DEVELOPMENT

We implemented a comprehensive "career enhancement" program that included:

- Basic AI literacy
- Advanced compliance analytics
- Leadership development
- Strategic planning skills

THE RESULTS

Despite some technical hurdles with data integration and initial resistance from senior staff, Southwest Healthcare achieved remarkable results within 12 months:

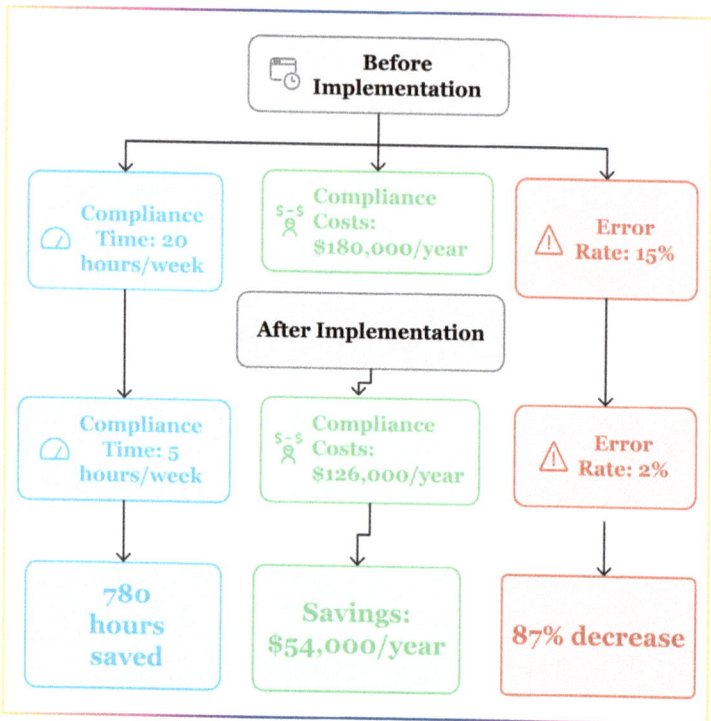

"*The journey wasn't always smooth,*" Sophia shared. "*We struggled for three weeks to integrate our legacy compliance systems, and our initial training approach needed significant adjustment. But working through these challenges actually helped us build a stronger program. Our team now comes to me with ideas for new AI applications. They're driving the innovation.*"

✓ REALITY CHECK

Myth: You need organization-wide buy-in before starting any AI implementation.

Reality: Starting small with clear wins in one department often creates organic buy-in across the organization.

Impact: Waiting for complete buy-in delays valuable benefits and can increase resistance.

AI Pro Tip: Begin with targeted implementations that demonstrate clear value and build momentum.

LESSONS FOR YOUR BUSINESS

Whether you're in healthcare or not, the principles we used at Southwest Healthcare and others apply broadly. Here's what you need to remember:

Start with Understanding

Take time to really listen to your team's concerns. Don't dismiss them as irrational; they're very real. When EndUp Furniture started its AI journey, Tom spent a full week having coffee with different team members, learning their specific concerns about automated collections. Those conversations shaped an implementation that achieved 85% team buy-in within the first month.

Show, Don't Tell

Demonstrate AI's benefits through small, successful pilots before rolling out larger initiatives. At Green Valley Landscaping, Nina started with just two maintenance routes. When those crews reported saving 90 minutes daily on drive time, other teams began asking to be next in line for route optimization.

Involve Your Team

Let them help shape how AI will be implemented in their areas of expertise. Duke's Mayonnaise saw this firsthand when they let their veteran production staff decide which quality control parameters to monitor first. Their insights led to a 30% reduction in quality variations within 90 days.

Focus on Enhancement

Consistently show how AI enhances capabilities rather than replaces them. TRAXX Flooring's customer service team initially feared automation would eliminate their roles. Instead, AI handling routine inquiries freed them to focus on complex customer needs and larger work orders, leading to increased sales commissions and a 45% increase in customer satisfaction scores.

COMMON PITFALLS TO AVOID

- Ignoring or dismissing employee concerns about AI
- Implementing AI without proper explanation or preparation
- Failing to demonstrate tangible benefits of AI adoption
- Neglecting to address ethical concerns related to AI
- Trying to force AI adoption without building a supportive culture

Chapter Summary: Remember what Sophia from Southwest Healthcare told me: *"The key wasn't forcing AI adoption—it was showing our team how AI made them more valuable."* Through their journey, we've seen how proper resistance management can transform skeptics into advocates.

KEY TAKEAWAYS:

- Address job displacement fears directly and early
- Start small and demonstrate clear wins
- Involve your team in implementation decisions
- Focus on enhancement rather than replacement
- Build a strong champion network

LOOKING AHEAD: FROM SUPPORT TO SUCCESS

As we conclude Part I and move into Part II of our journey, remember that overcoming resistance is just the beginning. While your team's support creates the essential foundation, now comes the million-dollar question that keeps business owners awake at night: How do you transform that support into measurable ROI and tangible business results?

I learned this lesson the hard way when a manufacturing client called me in desperation. Despite having an enthusiastic team and investing $85,000 in an advanced AI system, they were drowning in implementation chaos. "Everyone's excited about AI," their CEO told me, "but we're burning money without seeing results." Within weeks of applying a structured approach, they not only stopped the bleeding but generated a 215% return on their AI investment.

That client's expensive lesson reinforced what I've learned from implementing AI across 120+ businesses: Success doesn't come from enthusiasm or technology alone—it comes from following a proven path that consistently delivers results.

This brings us to Part II of our journey: The AI GROWTH Code. In the next chapters, I'll reveal the exact framework that's helped businesses like yours overcome operational roadblocks and achieve remarkable ROI—from Secured Tech's 85% reduction in support costs to Green Valley's 30% fuel savings. You'll learn not just what's possible with AI but exactly how to achieve these results in your business.

Remember: It's not about having the most advanced technology or the most enthusiastic team. It's about having the right approach that turns AI investments into business results. Let's discover your path to measurable success.

[1] https://www.microsoft.com/en-us/worklab/podcast/ stanford-professor-erik-brynjolfsson-on-how-ai-will-transform-productivity

Chapter 5: Winning Over the Skeptics (Yes, Even You)

PART TWO

THE AI GROWTH CODE

Now that you understand the real potential and limitations of AI for your business, it's time to turn that knowledge into action. The AI GROWTH Code is a proven path to AI success that I've refined through implementing AI in over 120 businesses.

In this section, you'll learn exactly how to:

- Transform business challenges into clear AI objectives
- Build your AI capability without breaking the bank
- Make AI work in your daily operations
- Scale your success across the organization

CHAPTER 6

INTRODUCTION TO THE AI GROWTH CODE

Every step in the AI GROWTH Code comes directly from businesses like yours that have successfully navigated this journey. You'll see how companies like NAR Warehouse used these exact steps to increase their effective storage capacity by 35% without expanding their footprint. You'll learn how Secured Tech Solutions cut their support response time from 4 hours to 3 minutes using this systematic approach.

This isn't theory - it's a practical guide built from real successes and hard lessons learned. Whether you're just starting with AI or looking to expand your current implementation, the AI GROWTH Code gives you a clear path forward.

Let's begin by understanding exactly how this proven framework can work in your business.

"Success leaves clues." In helping businesses implement AI, I've discovered something crucial: the difference between success and failure rarely comes down to the technology itself. It comes down to what we discussed in the introduction—having the right approach.

This insight is echoed by Matt McLarty, Chief Technology Officer at Boomi, who notes: "The biggest factor holding back AI implementation is people not knowing where to start. People need to understand that you don't actually have to be an expert on how to create generative AI in order to get value from generative AI." [MIT Technology Review, 2024]

He's right. I've seen firsthand how a clear framework matters more than technical expertise. Businesses that succeed with AI don't just adopt technology—they follow a systematic path that turns promising tools into proven results. This is exactly why I developed the AI GROWTH Code.

What began as a solution for a handful of clients who trusted me to help them navigate AI adoption has evolved into something much more powerful. Through each implementation, we tested, measured, and refined our approach. We kept and enhanced what worked and revised or replaced what didn't. The steps aren't theoretical—they're born from countless hours of solving real business challenges.

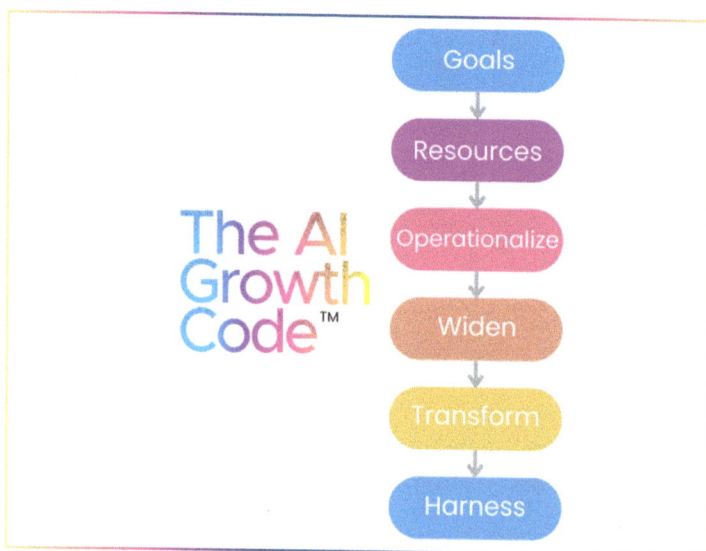

The AI Growth Code™

Goals
↓
Resources
↓
Operationalize
↓
Widen
↓
Transform
↓
Harness

Part Two: The AI GROWTH Code

The AI GROWTH Code transforms AI from a promising technology into a practical business advantage.

Remember those impressive ROI numbers we saw in Chapter 1— how businesses implementing AI solutions are achieving returns ranging from 132% to 353%? Those results don't come from random implementation. They come from following a proven path. Just as importantly, they come from avoiding the costly mistakes we discussed earlier—like the manufacturing company that wasted $78,000 on advanced AI tools they weren't ready to use.

Let me show you how this framework works in the real world. When Frank from NAR Warehouse first reached out, his situation might sound familiar. Managing a 200,000-square-foot distribution center with over 200 employees, he faced mounting challenges. *"Rich, I know AI could help us be more efficient, but where do we even start? We're wasting time picking errors, and our space utilization is a mess."*

His numbers told a stark story: a 4% error rate in order picking was costing them $380,000 annually in rework and customer satisfaction issues. Their manual space allocation methods meant that 27% of their warehouse capacity was consistently underutilized, while peak seasons brought chaos and missed deliveries.

"Everyone was selling us different AI solutions," Frank recalled during our first meeting. *"Some wanted us to do a complete digital transformation. Others said we needed machine learning and data science teams. It was paralyzing."* Like many business leaders we discussed in Chapter 4, Frank wasn't resistant to change—he was resistant to chaos.

Today, NAR Warehouse operates with 99.9% picking accuracy, has increased its effective storage capacity by a third without expanding its footprint, and handles 22% more orders with the same staff. Their success didn't come from choosing the most expensive AI solution or the most advanced technology. It came from following a systematic approach to implementation—exactly the approach separating 29% of businesses seeing increased revenue with AI from those still struggling to get started.

Let me show you exactly how they did it and how you can achieve similar results in your business...

NAR's transformation through the GROWTH framework demonstrates exactly how systematic implementation turns AI from a promising technology into practical results:

GOALS: GETTING CLARITY FIRST

"The hardest part wasn't starting," Frank admitted. *"It was choosing where to start."* Like many businesses we've discussed, NAR had multiple opportunities for AI implementation. Their initial instinct was to try solving everything simultaneously: warehouse management, order picking, inventory control, and space utilization all seemed equally urgent.

Instead, we used the Impact/Effort Matrix, which I'll introduce you to in Chapter 8, to identify their highest-value starting point. After analyzing their operations, one opportunity stood out: order-picking accuracy. Not only was it costing them $380,000 annually in direct losses, but it affected everything from customer satisfaction to staff morale.

We set three clear, measurable objectives:

- Reduce picking errors by 75% within 90 days
- Cut rework time by 50%
- Improve customer satisfaction scores by 25%

"Having specific targets changed everything," Frank explained. *"Instead of trying to boil the ocean, we could focus our energy where it mattered most."*

RESOURCES: BUILDING THE FOUNDATION

With clear goals established, we conducted a comprehensive resource assessment. This revealed something surprising: NAR already had 60% of the needed technology. Their warehouse management system included AI capabilities they'd never activated. The real resource gaps weren't in technology but in training and processes.

We developed a three-part resource plan:

1. Activate and optimize existing AI capabilities
2. Train team leads as AI implementation champions
3. Create clear processes for data collection and analysis

"We thought we needed to buy new everything," Frank noted. *"We needed to better use what we already had."*

OPERATIONALIZE: MAKING IT REAL

Implementation began with a single picking zone—the "prove it" area. Rather than disrupting the entire warehouse, we chose a controlled environment where we could demonstrate success and refine our approach.

The initial results exceeded expectations:

- Picking errors dropped to 0.3% within two weeks
- Processing time decreased by 45%

"The numbers were impressive," Frank shared, *"but what sold everyone was how much easier their jobs became.* Our pickers went from constantly solving problems to consistently preventing them."

WIDEN: EXPANDING SUCCESS

With proven success in the pilot zone, expansion followed naturally. We developed a systematic roll-out plan that covered one zone at a time, incorporating lessons learned at each step. This measured approach allowed us to:

- Refine training procedures
- Adjust workflows based on feedback
- Address challenges before they become problems
- Build confidence throughout the organization

"Each new zone implemented better than the last," Frank explained. *"We weren't just copying what worked—we were improving it."*

TRANSFORM: RE-IMAGINING OPERATIONS

As AI became integrated into daily operations, something remarkable happened. Teams started identifying new opportunities for improvement. The data from order picking helped optimize inventory placement. Better inventory placement improved space utilization, leading to more efficient operations.

Within six months, NAR had:

- Increased effective storage capacity by 31%
- Reduced walking time for pickers by 40%
- Improved overall warehouse efficiency by 42%

HARNESS: DRIVING CONTINUOUS IMPROVEMENT

Today, NAR uses its AI capabilities to drive continuous improvement. Their system doesn't just track errors—it predicts and prevents them. It doesn't just optimize current operations—it identifies future opportunities.

"The biggest change," Frank reflected, *"isn't in our numbers, though they're impressive. It's in how we think about our business. We've gone from reacting to predicting, fixing to preventing, hoping to knowing."*

The results speak for themselves:

- 99.9% picking accuracy
- 22% more orders processed
- 31% increase in storage utilization
- 85% reduction in customer complaints
- Zero staff reductions

BUT WHAT ABOUT MY BUSINESS?

You might be asking yourself, *"But Rich, my business isn't a warehouse."* That's exactly what Laura from a local flower shop told me and what Ronnie, who makes and sells some of the best Southern BBQ sauce around, said, too. I tell every business owner I work with: The AI GROWTH Code works across any industry because it focuses on process, not just technology.

While your journey might not follow the AI GROWTH Code exactly as NAR did, having this framework ensures you're making informed decisions rather than reactive ones. Too many businesses waste time and money trying to 'wing it' with AI. Successful companies follow a structured approach, even if they need to adjust it along the way.

Let me show you how different businesses adapted the GROWTH framework to their specific needs:

Laura's Flower Shop: Small Business, Big Impact

When Laura first approached me, she was drowning in order management during peak seasons like Valentine's Day and Mother's Day. *"I'm not running a warehouse,"* she said, *"I'm just trying to keep up with orders and keep flowers fresh."*

Using the GROWTH framework, we:

- Started with the specific goal of reducing order processing time
- Leveraged basic AI tools she already had access to
- Implemented a simple customer communication system
- Expanded to inventory management
- Transformed her entire ordering process

Result: Order processing time dropped from 15 to 2 minutes per order, and flower waste decreased by 25%.

Ronnie's BBQ Sauce: From Chaos to Control

Ronnie's challenge was different, predicting demand for his artisan BBQ sauces across multiple sales channels. *"I'm not Amazon,"* he told me. *"I just need to know how much sauce to make and when."*

His GROWTH journey focused on:

- Setting clear inventory management goals
- Using AI to analyze sales patterns
- Implementing predictive ordering
- Expanding to multiple retail channels
- Transforming his production planning

Result: Stock-outs were reduced by 85%, and revenue increased by 37% through better inventory management.

Key Learning: Adapt, Don't Copy

The power of the GROWTH framework isn't in copying what worked for others; it's in adapting the principles to your specific needs. What Laura, Ronnie, and dozens of other business owners discovered was that while their businesses were different from NAR Warehouse, the fundamental approach to implementing AI remained the same:

1. Start with clear business objectives
2. Build on existing resources
3. Focus on systematic implementation
4. Scale what works
5. Transform processes, not just automate them

KEY TAKEAWAYS FROM THE AI GROWTH CODE

SUCCESS PRINCIPLES:

- Every successful AI implementation starts with a plan, not just technology
- Each component builds naturally on the previous one
- Starting small and scaling with success prevents costly mistakes
- The framework adapts to any industry or business size
- Real results come from systematic implementation

THE AI GROWTH CODE IN ACTION:

GOALS:

- Average 35% operational improvement achieved by starting with clear metrics
- Quick Wins build momentum for larger projects
- Measurable targets enable progress tracking
- Clear objectives reduce resistance

RESOURCES:

- Typical ROI of 150-300% within 180 days when properly resourced
- Efficient resource allocation trumps total investment size
- Start with existing capabilities where possible
- Build foundations before expanding

OPERATIONALIZE:

- 85% of successful implementations start with small, focused projects
- Quick Win achievements create natural momentum
- Team buy-in grows through visible success
- Clear processes enable consistent results

WIDEN:

- 75-85% successful expansion rate when following proven paths
- 30-45% additional cost savings through systematic growth
- 25-35% efficiency gains in new areas
- Methodical expansion reduces risks

TRANSFORM:

- 45% average cost reduction through full transformation
- Process improvements compound over time
- Cultural evolution drives lasting change
- The systematic approach ensures sustainability

HARNESS:

- 25-45% additional value creation through optimization
- Sustainable competitive advantages emerge
- Continuous improvement culture develops
- Long-term success patterns establish

COMMON PITFALLS TO AVOID:

- Trying to skip steps in the framework
- Attempting too many changes at once
- Not involving your team early enough
- Focusing on technology before strategy
- Rushing through the planning phase
- Not measuring progress at each step

LOOKING AHEAD

In the next chapter, we'll dive deep into the 'G' of GROWTH - Goals. This is where your AI journey begins to take shape. You'll learn exactly how Frank from NAR Warehouse, Laura from the flower shop, and dozens of other business owners set clear, actionable goals that drove their success. Whether you're looking to improve customer service, streamline operations, or boost sales, having the right goals will determine your success with AI.

Remember: The journey of a thousand miles begins with a single step.

Are you ready to take the first one?

CHAPTER 7

GOALS – DEFINE PRECISE BUSINESS OBJECTIVES

The AI
Growth
Code™

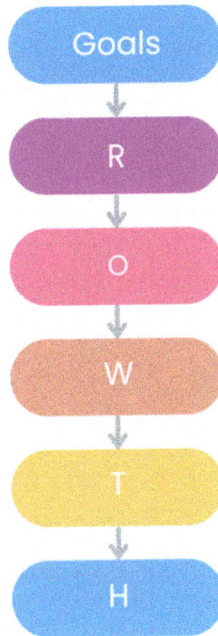

The 'G' in The GROWTH Code stands for Goals. The essential first step that determines the success of every AI implementation I've overseen. Over my career, I've learned that success doesn't start with technology; it starts with clarity of purpose.

"Where should we start with AI?" is the first question almost every business owner asks me. My answer always surprises them: *"Let's not talk about AI yet. Let's talk about what keeps you up at night."*

When Tom from EndUp Furniture and I first spoke, he had already researched dozens of AI tools. *"Rich, which one should we implement first?"* he asked. I responded with another question: *"What's your biggest business challenge right now?"*

His answer was immediate: *"Late payments. We're spending too much time chasing money already earned, and it's killing our cash flow."*

Now we had something real to work with - not just a vague desire to *"use AI,"* but a specific business challenge that needed solving. This is where the goal-setting process begins and why it forms the foundation of the AI GROWTH Code. Without this clarity, the remaining components - Resources, Operationalize, Widen, Transform, and Harness - lack direction and purpose.

This experience leads us to two essential questions that will help you identify the right starting point:

1.   What's Keeping You Up at Night?

Identify your most pressing business challenges.

2.   What Does Success Look Like?

Define clear, measurable objectives using the well-known SMART framework.

✓ REALITY CHECK

Myth: You need comprehensive AI knowledge before setting goals.

Reality: Start with business problems, not technology solutions.

Impact: Technology-first thinking often leads to wasted investments.

AI Pro Tip: Begin with your most pressing business challenge.

SETTING SMART GOALS

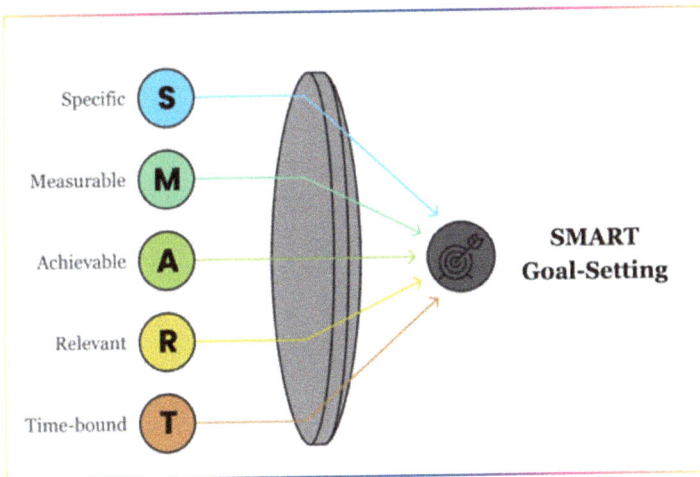

Let's examine these questions using Tom's experience as our guide.

1. WHAT WAS KEEPING TOM UP AT NIGHT?

This question seems simple, but it requires brutal honesty. When Tom first identified late payments as his biggest challenge, we dug deeper to understand the full impact:

- How many accounts were typically overdue?
- How much time was spent chasing payments?
- What was the effect on cash flow?
- How was this affecting his team?

This detailed understanding would prove crucial for the next step.

2. WHAT DOES SUCCESS LOOK LIKE FOR TOM?

This is where we employ the SMART framework to turn a business challenge into a clear, actionable goal. For Tom, this meant:

Specific:

- Bad: "Fix our collections problem"
- Better: "Reduce late payments using AI"
- Best: "Use AI to automate payment reminders and predict payment behavior for accounts over 30 days late"

Measurable:

- Bad: "Collect payments faster"
- Better: "Reduce the number of late payments"
- Best: "Reduce accounts over 30 days late from 35% to 15%, representing $42,000 in monthly cash flow improvement"

Achievable:

- Bad: "Eliminate all late payments"
- Better: "Cut late payments in half"

- Best: "Reduce late payments by 50% based on historical payment patterns and current AI capabilities"

Relevant:

- Bad: "Because AI seems useful"
- Better: "To improve cash flow"
- Best: "To improve cash flow while maintaining strong customer relationships and freeing up 3 hours daily of collection team time"

Time-bound:

- Bad: "As quickly as possible"
- Better: "This quarter"
- Best: "Within 90 days of AI implementation, with 25% improvement by day 30"

Final SMART Goal: "Implement AI payment prediction and automation system to reduce accounts over 30 days late from 35% to 15% within 90 days, improving monthly cash flow by $42,000 while maintaining customer satisfaction above 4.5/5 and reducing collection team time by 3 hours daily."

"Breaking down each element of SMART helped us see opportunities we were missing," Tom shared. *"For example, adding the customer satisfaction metric kept us focused on maintaining relationships while improving collections."*

Now, let's break down a common AI implementation goal and see how each element of SMART makes it stronger:

Initial Goal: "We want to use AI to help with customer service"

Specific:

- Bad: "Improve customer service"
- Better: "Implement AI chatbot to handle basic customer inquiries"

- Best: "Deploy AI chatbot to handle top 20 most common customer questions"

Measurable:
- Bad: "Reduce customer wait times"
- Better: "Cut response time to under 5 minutes"
- Best: "Reduce average initial response time from 4 hours to under 5 minutes"

Achievable:
- Bad: "Handle all customer service with AI"
- Better: "Automate 50% of customer inquiries"
- Best: "Automate 50% of customer inquiries based on the current volume of 200 daily questions"

Relevant:
- Bad: "Because everyone else is using AI"
- Better: "To improve customer satisfaction"
- Best: "To improve customer satisfaction while reducing support team workload by 30%"

Time-bound:
- Bad: "As soon as possible"
- Better: "Within six months"
- Best: "Within 90 days of implementation"

Final SMART Goal: "Implement AI a chatbot to handle the top 20 most common customer inquiries (representing 50% of current 200 daily questions), reducing average response time from 4 hours to under 5 minutes while decreasing support team workload by 30%, completed within 90 days of implementation."

Let's look at another example with Green Valley Landscaping's route optimization goal:

GREEN VALLEY LANDSCAPING'S SMART GOAL EVOLUTION

When Nina first discussed using AI, her initial goal was to *"stop wasting fuel on inefficient routes."* Let's see how we refined this using each SMART element:

Specific:

- Bad: "Make our routes better"
- Better: "Optimize routes to use less fuel"
- Best: "Use AI to create optimized routes for 25 maintenance crews across our three service zones, factoring in traffic patterns, job duration, and crew specialties"

Measurable:

- Bad: "Reduce fuel usage"
- Better: "Cut monthly fuel costs"
- Best: "Reduce monthly fuel costs from $12,000 to $8,400 while increasing completed jobs from 8 to 10 per crew per day"

Achievable:

- Bad: "Eliminate all wasted drive time"
- Better: "Reduce drive time by half"
- Best: "Reduce drive time by 30% based on current service area analysis and proven AI routing capabilities"

Relevant:

- Bad: "Because fuel costs are rising"
- Better: "To improve profitability"
- Best: "To increase profitability while improving crew satisfaction and on-time arrival rates for customers"

Time-bound:

- Bad: "Before peak season"
- Better: "Within six months"
- Best: "Full implementation within 120 days, with the first zone optimized within 30 days"

Final SMART Goal: "Implement AI route optimization for 25 maintenance crews to reduce monthly fuel costs by 30% (from $12,000 to $8,400) while increasing daily completed jobs from 8 to 9 per crew and achieving 95% on-time arrival rate, with first zone completed in 30 days and full implementation within 120 days."

"The SMART framework helped us see this wasn't just about saving fuel," Nina explained. *"By being specific about crew productivity and customer satisfaction, we built a much stronger business case for AI."*

Let's examine one more example of Bon Secours Hospital's hiring process goal:

BON SECOURS HOSPITAL'S SMART GOAL EVOLUTION

Olivia's initial goal was to *"speed up our recruitment process."* Here's how we refined this using each SMART element:

Specific:

- Bad: "Make hiring faster"
- Better: "Reduce time-to-hire with AI"
- Best: "Use AI to screen candidates, automate initial interviews, and predict top performers for our 15 most critical nursing positions while maintaining compliance with healthcare regulations"

Measurable:

- Bad: "Fill positions quicker"
- Better: "Cut hiring timeline in half"
- Best: "Reduce time-to-hire from 45 days to 21 days while improving first-year retention rate from 75% to 85% and cutting recruitment costs from $65,000 to $35,000 annually"

Achievable:

- Bad: "Automate all hiring decisions"
- Better: "Automate candidate screening"
- Best: "Automate 80% of initial candidate screening and 60% of preliminary interviews based on current AI capabilities and regulatory requirements"

Relevant:

- Bad: "Because manual screening takes too long"
- Better: "To improve hiring efficiency"
- Best: "To reduce critical position vacancies, improve quality of hire, and free up HR team for strategic tasks while maintaining compliance and candidate experience"

Time-bound:

- Bad: "As soon as it's working"
- Better: "By next quarter"
- Best: "System implemented and staff trained within 90 days, with first automated screening live in 30 days"

Final SMART Goal: "Deploy AI hiring system to reduce time-to-hire from 45 to 21 days for critical nursing positions, automate 80% of initial screening, improve first-year retention to 85%, and reduce annual recruitment costs by $30,000 while maintaining 100% compliance. Phase 1 screening automation is to go live in 30 days, with full implementation in 90 days."

'*The SMART process revealed opportunities we hadn't considered,*' Olivia shared. '*Breaking down each component helped us create clear milestones and measurable outcomes that directly tied to our hospital's needs.*'"

These examples show how the SMART framework transforms vague intentions into actionable goals. The process works across any industry or business challenge, but success depends on following a systematic approach.

Let me share the specific steps I use with my clients to create effective SMART goals:

Start With Your Pain Point

At EndUp Furniture, Tom's team was drowning in late payments. Rather than jumping to solutions, we first documented:

- Number of overdue accounts
- Average days outstanding
- Time spent on collections
- Impact on cash flow
- Effect on team morale

Quantify Current State

Nina at Green Valley knew routes were inefficient, but we needed specifics:

- Current fuel costs
- Time spent planning routes
- Number of jobs completed
- Customer satisfaction scores
- Team overtime hours

Define Ideal Outcomes

For Bon Secours, Olivia identified clear targets:

- Specific reduction in hiring time
- Improvement in retention rates
- Decrease in recruitment costs
- Compliance maintenance

Set Realistic Time-frames

Each business sets achievable timelines based on its resources:

- EndUp: 30 days for initial implementation
- Green Valley: 120 days for the full rollout
- Bon Secours: 90 days for complete system deployment

Create Clear Metrics

Success metrics varied by business but always included:

- Primary goal achievement
- Secondary benefits
- Resource utilization
- Implementation milestones

"'Having these clear steps made goal-setting precise and measurable,' Tom shared. *'Instead of vague objectives, we had specific targets we could track and measure success.'"*

Key Takeaways from Goal Setting:

Through helping hundreds of businesses implement AI, I've learned that successful goal-setting comes down to these essential elements:

Focus on Business Challenges First

- Start with what keeps you up at night
- Document specific pain points

- Quantify current impact
- Understand downstream effects

Use SMART Framework Consistently

- Transform vague ideas into specific targets
- Create measurable objectives
- Set realistic time-frames
- Ensure business relevance

Build on Success

- Start with one clear goal
- Document your results
- Learn from the process
- Expand thoughtfully

COMMON GOAL-SETTING PITFALLS TO AVOID

Through helping over 120 businesses set AI goals, I've seen how seemingly minor mistakes in goal-setting can derail entire implementations. Let me share the most expensive lessons my clients have learned.

The "Everything, Everywhere, All at Once" Trap

A manufacturing client wanted to automate their entire production line, customer service, and logistics simultaneously. *"We thought bigger meant better,"* their COO admitted. After three months of chaos and $45,000 wasted, we refocused on one specific goal: reducing quality control errors by 35%. They achieved this in 60 days, creating momentum for further improvements.

The "Impossible Timeline" Mistake

"We can roll this out company-wide in two weeks," an enthusiastic retail manager told me. His team burned out trying to meet

an unrealistic deadline. When we reset with a 90-day phased implementation, they actually completed their AI customer service rollout ahead of schedule.

The "Fuzzy Metrics" Problem

A distribution company set a goal to *"significantly improve efficiency."* Without specific metrics, they couldn't prove success or justify further investment. When we redefined their goal to *"reduce order processing time from 45 to 15 minutes,"* they achieved it within 75 days and secured funding for expansion.

The "Technology First" Error *"We started by choosing the AI tool, then figured out what to do with it,"* a services firm leader shared after wasting $28,000. When we reversed the process - starting with clear business objectives - they chose the right tool and saw ROI within 45 days.

The "Copy and Paste" Mistake

After hearing about a competitor's success with AI-driven inventory management, a retail chain tried to replicate their exact goals. However, it forgot that their business model, customer base, and challenges were different. Resetting its goals to match its specific needs led to 25% better inventory accuracy.

✓ REALITY CHECK

Myth: Setting ambitious goals shows leadership vision.

Reality: Setting achievable goals builds momentum and trust.

Impact: Unrealistic goals often lead to abandoned AI projects.

AI Pro Tip: Start with one specific, measurable goal that can be achieved within 90 days.

TURN GOALS INTO ACTION: YOUR AI PLANNING TOOL

To help you translate your AI ambitions into concrete, achievable goals and communicate their benefits to various stakeholders, I've developed two tools to assist you:

Tool #3 - SMART Goals Worksheet

The SMART Goals Worksheet eliminates the guesswork of setting goals for AI initiatives. Built from real implementation experiences, this comprehensive system includes validation criteria and a library of successful examples you can adapt for your business. I've watched this structured approach double project success rates by ensuring every goal is specific, measurable, achievable, relevant, and time-bound.

Tool # 4 - AI Stakeholder Impact Analysis and Communication Toolkit

The AI Stakeholder Impact Analysis and Communication Toolkit helps businesses map stakeholders, anticipate their concerns, and create targeted engagement strategies to increase stakeholder buy-in from all levels of your organization.

KEY TAKEAWAYS: SETTING GOALS THAT DRIVE RESULTS

Goal-Setting Framework Success Patterns:

- Start with business problems, not AI solutions
- Focus on measurable outcomes
- Create clear implementation timelines
- Build team alignment from the start

SMART Goals That Work:

- Specific: Target one clear business problem
- Measurable: Include concrete numbers and metrics
- Achievable: Base targets on proven case studies
- Relevant: Connect directly to business value
- Time-bound: Set realistic implementation schedules

Real-World Impact:

- EndUp Furniture: 50% reduction in late payments in 90 days
- Green Valley: 30% fuel cost reduction in 120 days
- Bon Secours: Hiring time cut in half within 90 days
- Duke's Mayonnaise: Quality variation reduced by 35% in the first quarter

Priority Assessment Indicators:

- Clear connection to the bottom line
- Available data to measure success
- Defined implementation path
- Team capability to execute
- Resources ready for deployment

Common Goal-Setting Mistakes to Avoid:

- Setting vague or unmeasurable goals
- Trying to solve everything at once
- Skipping the quantification step
- Setting unrealistic timeframes
- Not involving key stakeholders

LOOKING AHEAD

Clear goals provide the foundation for everything that follows in your AI journey. However, even the best objectives need the right resources to become a reality.

This brings us to the 'R' in our GROWTH Code - Resources. In Chapter 8, we'll explore this critical component in two parts:

First, we'll discover how to build your AI Infrastructure - the complete foundation of people, technology, processes, and resources that make AI work in your business. You'll learn how successful companies built foundations that saved money and accelerated their success.

Then, we'll explore matching your business needs to the right AI solutions. Just as a master craftsman chooses the right tools for specific jobs, you'll learn how to select and implement AI capabilities that align with your goals and infrastructure.

Remember Tom's insight from EndUp Furniture: *"Our success with AI didn't come from choosing the most expensive tools—it came from building the right foundation first."*

Let's build yours.

CHAPTER 8

RESOURCES—PARTS ONE & TWO

The AI Growth Code™

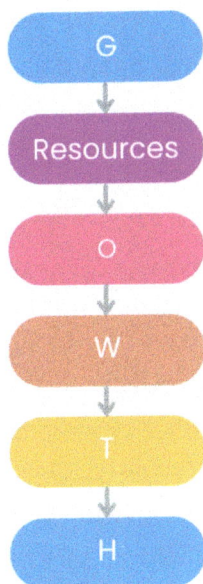

G

Resources

O

W

T

H

RESOURCES (PART ONE) – BUILDING YOUR AI INFRASTRUCTURE

" **W**e almost wasted $85,000 on infrastructure we didn't need," Mitch, with Cornerstone Construction, shared during our first meeting. "*Instead, we spent $47,000 on exactly what would drive results, and we were live in 90 days instead of 180.*"

Mitch's first question reflected a challenge I hear almost daily: "*Rich, I understand our goals, but how do we make this happen? As builders, we know a structure is only as good as its foundation, but what infrastructure do we need for AI?*"

His construction background provided the perfect analogy. Through implementing AI across various industries, I've learned that success depends on building the right AI Infrastructure—the complete foundation of people, technology, processes, and resources that make AI work in your Business.

Think of AI Infrastructure like building a house: you need not just the materials and tools but also the right workers, proper permits, and solid ground to build on.

Let me show you how companies like Cornerstone Construction, Secured Tech Solutions, and EndUp Furniture built their AI Infrastructure from the ground up. Their experiences will give you a practical blueprint for evaluating and strengthening your foundation.

"*I thought having the right software was enough,*" Mitch admitted during our initial consultation. "*You showed us that AI Infrastructure needs to be built systematically, like constructing a building - one solid layer at a time.*"

Myth: You need to rebuild your entire technical infrastructure for AI.

Reality: Most businesses can leverage existing systems as a foundation.

Impact: Complete rebuilds waste valuable existing infrastructure.

AI Pro Tip: Use the Assessment Tool to identify what you can build upon.

UNDERSTANDING THE FOUR PILLARS OF AI INFRASTRUCTURE

Your AI Infrastructure consists of four essential pillars, each contributing 25% to your overall readiness. Like the classical columns that have supported important structures for thousands of years, these pillars work together to create a stable platform for your AI success:

1. People & Skills - Your human foundation of talent, expertise, and adaptability
2. Budget & Resources - Your financial foundation that fuels implementation and growth
3. Implementation Readiness - Your operational foundation ensures smooth integration
4. Technology & Tools - Your technical foundation that delivers capabilities

As Mitch discovered at Cornerstone Construction, the entire structure becomes unstable if any single pillar is weak or missing.

But when properly built and aligned, these four pillars create the solid foundation you need to succeed with AI.

Let's examine each pillar in detail, starting with the most crucial element of any successful AI initiative - your people.

I've learned that getting these pillars right from the start typically reduces total implementation costs by 35-40% and cuts launch time by half.

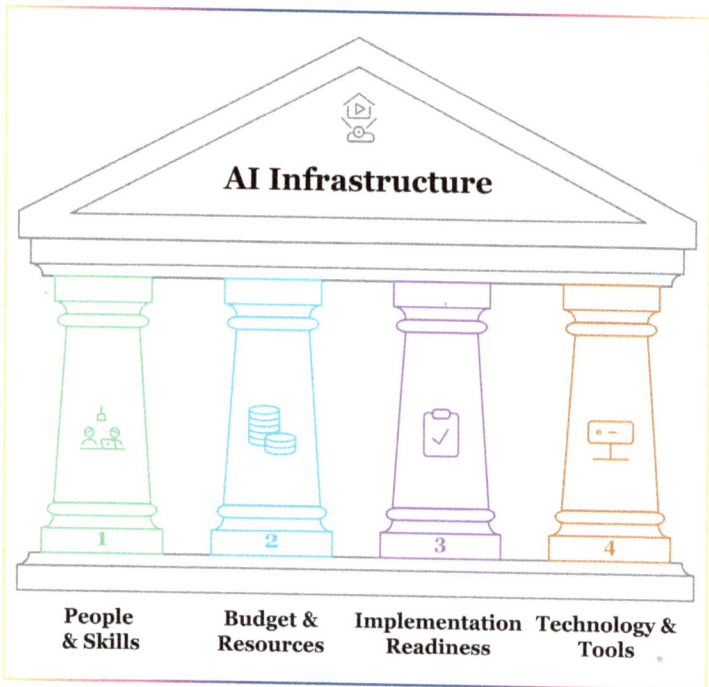

AI Infrastructure

1 People & Skills
2 Budget & Resources
3 Implementation Readiness
4 Technology & Tools

PILLAR 1: PEOPLE & SKILLS – YOUR HUMAN FOUNDATION

This Section will teach you how to assess your team's capabilities, develop necessary skills, and build a culture ready for AI adoption. We'll explore training ("career enhancements") frameworks, change management strategies, and success metrics.

People & Skills = Current Skills + Training Needs + Leadership Capabilities + Roles to Fill

CASE STUDY: SECURED TECH SOLUTIONS

"Our biggest challenge wasn't technical - it was getting our experienced support team to trust AI," Joshua explained. *"We achieved our 95% adoption rate by starting with what they already knew."*

Let me show you how they built their human foundation systematically across four key areas:

Current Skills Assessment When Joshua's team began their AI journey, they started by documenting exactly where they stood:

- Only 4 out of 10 team members were comfortable with digital tools
- Key gaps in understanding basic automation concepts
- Strong foundation in support ticketing software
- Need for basic AI awareness training

"Taking inventory of our capabilities wasn't just about finding gaps," Joshua shared. *"It helped us discover strengths we didn't know we had."*

Training Needs Analysis With a clear picture of their current capabilities, Secured Tech Solutions developed a structured approach:

- Created micro-learning modules (15-20 minutes each)
- Established peer learning networks
- Developed hands-on practice scenarios
- Built feedback loops for continuous improvement
- Results: Reduced training time by two/thirds while improving knowledge retention.

"The key was making learning part of daily work," Joshua noted. *"Instead of long training sessions, we integrated micro-learning into regular operations."*

Roles & Responsibilities They identified and developed three critical roles:

- AI Champions: Tech-savvy team members who could bridge the gap between technology and daily operations
- Process Owners: Experienced Staff who understood which workflows needed enhancement
- Change Agents: Natural leaders who could help drive adoption

Their approach:

- Started with AI handling routine tasks, ones the team disliked (saved 15 hours/week)
- Used early wins to build confidence (response time improved by 45% in the first week)
- Created peer mentoring pairs between tech-savvy and experienced Staff
- Built progression road maps showing how AI would enhance careers, not replace jobs

Leadership Capabilities Joshua's team discovered that leadership development was crucial for sustainable success:

- The executive team showed a strong commitment to innovation
- Middle management needed more AI exposure
- Clear decision-making processes established
- Change management protocols implemented

"After 60 days, the same people who resisted AI were suggesting new ways to use it," Joshua observed. *"Our change resistance dropped from 6 out of 10 to 2 out of 10 when people saw how it made their jobs better, not obsolete."*

Results After 90 Days:

- Team adoption rates doubled
- Technical competency scores improved by 85%
- Change resistance decreased by 66%

INFRASTRUCTURE INVESTMENT GUIDELINES BY BUSINESS SIZE

(Note: These costs reflect only the people & skills foundation - AI tool costs will be covered in Part Two of this chapter)

Small Business (1-10 employees)

Typical Infrastructure Investment: $2,400-6,000/year

- Training & skill development programs: 60%
- External expertise/consulting: 40%
- Typical ROI time frame: 30-45 days
- What's included: Team training, change management support, expertise development

- What's not included: Actual AI tools and software costs (covered in Part Two)

Medium Business (11-50 employees)

Typical Infrastructure Investment: $6,000-24,000/year

- Training programs: 40%
- Change management: 30%
- External expertise: 30%
- Typical ROI time-frame: 45-60 days
- What's included: Team development programs, process optimization, and management training
- What's not included: AI platform costs and implementation fees (detailed in Part Two)

Large Business (50+ employees)

Typical Infrastructure Investment: $24,000-60,000/year

- Enterprise training: 35%
- Change management: 35%
- External expertise: 30%
- Typical ROI time frame: 60-90 days
- What's included: Enterprise-wide training, change management programs, expert guidance
- What's not included: Enterprise AI solutions and custom development (explored in Part Two)

Implementation Support

While some larger organizations build these foundations internally, The AI Pros team can provide training, change management, and expertise development support at each level. Our experience implementing these foundations across 120+ businesses helps reduce typical infrastructure build time by 40-50%.

Learn more about implementation support options at AIGrowthCode.com

---

✓ REALITY CHECK

Myth: Technical expertise is the most important people component.

Reality: Change adaptability and learning capacity matter more.

Impact: Many companies waste resources on technical training before building change capability.

AI Pro Tip: Start with the change readiness assessment before technical training.

PILLAR 2: BUDGET & RESOURCES - YOUR FINANCIAL FOUNDATION

In this Section, you will discover how to allocate resources effectively, plan for implementation costs, and ensure sustainable funding for your AI initiatives. We'll examine cost structures, ROI calculations, and resource optimization strategies.

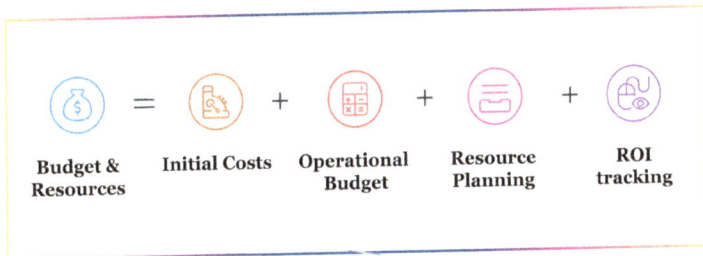

| Budget & Resources | = | Initial Costs | + | Operational Budget | + | Resource Planning | + | ROI tracking |

CASE STUDY: ENDUP FURNITURE

Tom's initial focus on just software costs almost derailed their implementation. *"We budgeted $85,000 for AI tools,"* Tom shared, *"but our comprehensive assessment revealed we could achieve better results with $47,000 by investing strategically."*

Let me show you how they built their financial foundation across four key areas:

Initial Costs

EndUp developed a clear understanding of implementation costs:

- Software licensing and platform fees
- Necessary system upgrades
- Initial training programs
- Implementation support

"The biggest mistake would have been buying expensive tools without understanding our complete needs," Tom explained. *"By mapping out all initial costs, we avoided expensive surprises."*

Operational Budget

They created a sustainable funding structure:

- Monthly subscription management
- Ongoing maintenance costs
- Regular updates and upgrades
- Support services

Results: Reduced ongoing costs by 35% through strategic planning and eliminated unnecessary expenses.

Resource Planning

EndUp's focused approach to resource allocation:

- Team time commitments
- Training requirements
- Implementation scheduling
- Support needs

"We learned to treat time as carefully as money," Tom noted. *"Understanding the full resource picture helped us implement efficiently."*

ROI Tracking

They established clear metrics:

- Implementation cost savings
- Time to value measurement
- Productivity improvements
- Operational efficiencies

Results After 90 Days:

- Implementation costs 45% below initial estimates
- ROI achieved 40% faster than the industry average
- Resource utilization improved by 65%
- Break-even achieved in 41 days

✓ REALITY CHECK

Myth: AI costs are just about buying software.

Reality: A complete AI Infrastructure requires investment across multiple areas.

Impact: Underestimating full resource needs leads to stalled implementations.

AI Pro Tip: Use the Assessment Tool to plan your complete financial foundation.

BUDGET & RESOURCES INFRASTRUCTURE CONSIDERATIONS

(Note: Most direct costs associated with AI will be covered in Part Two of this chapter)

The primary investments in this pillar typically involve:

System Evaluation

- Assessment of current systems' AI readiness
- Identification of necessary upgrades
- Integration capability review

Potential Infrastructure Upgrades Small Business Example:

- Storage capacity expansion: $200-600
- Memory upgrades: $100-300
- Network improvements: $300-900

(Only if current systems don't meet minimum requirements)

Resource Planning Time Investment

- 2-5 hours for initial assessment
- 1-2 hours weekly for monitoring
- Quarterly review meetings

PILLAR 3: IMPLEMENTATION READINESS – YOUR OPERATIONAL INFRASTRUCTURE

This Section teaches you how to prepare your organization for successful AI integration. We'll explore workflow assessment, communication strategies, documentation needs, and performance monitoring.

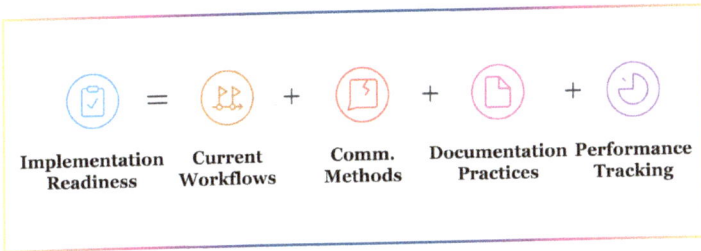

Implementation Readiness = Current Workflows + Comm. Methods + Documentation Practices + Performance Tracking

CASE STUDY: SECURED TECH SOLUTIONS

"We needed a systematic way to bring everything together," Joshua explained. *"Without proper implementation readiness, we risked building on shaky ground."*

Let me show you how they built their operational foundation across four key areas:

Current Workflows Secured Tech began with a thorough analysis of existing processes:

- Mapped critical workflows
- Identified bottlenecks
- Documented decision points
- Assessed integration opportunities

"Understanding our current processes helped us see exactly where AI could make the biggest difference," Joshua shared. *"It prevented us from disrupting what was already working well."*

Communication Methods They established clear channels for:

- Team updates
- Cross-department coordination
- Training delivery
- Support requests

Chapter 8: Resources—Parts One & Two

Results: Response times improved by 70%, and team alignment increased by 85%.

Documentation Practices The team created comprehensive resources for:

- Process guides
- Training materials
- Integration procedures
- Troubleshooting protocols

"Good documentation turned individual knowledge into organizational capability," Joshua noted. *"It helped us scale successfully across our entire support operation."*

Performance Tracking They implemented robust monitoring:

- System performance metrics
- Team adoption rates
- Customer satisfaction scores
- ROI measurements

Results After 90 Days:

- Team adoption reached 90%
- Support response time dropped from 4 hours to 3 minutes
- Customer satisfaction increased to 4.5/5 over 90 days
- Process efficiency improved by 85%

IMPLEMENTATION READINESS INFRASTRUCTURE CONSIDERATIONS

(Note: Direct implementation costs for AI tools will be covered in Part Two of this chapter)

Time Investment by Business Size:

Small Business (1-10 employees):

- Process documentation: 4-6 hours initially
- Workflow mapping: 2-3 hours total
- Readiness assessment: 2 hours
- Team preparation: 1 hour weekly
- Total time investment: 8-12 hours in the first month

Medium Business (11-50 employees):

- Process documentation: 8-12 hours initially
- Workflow mapping: 2-3 hours per department
- Readiness assessment: 3-4 hours
- Team preparation: 2 hours weekly
- Total time investment: 15-25 hours in the first month

Large Business (50+ employees):

- Process documentation: 15-20 hours initially
- Workflow mapping: 3-4 hours per department
- Readiness assessment: 4-6 hours
- Team preparation: 3-4 hours weekly
- Total time investment: 30-40 hours in the first month

When Additional Tools Might Be Needed:

Complex Process Documentation

- Base: Microsoft Office/Google Workspace (typically already owned)
- When to upgrade: Multiple departments need simultaneous access
- Recommended tool: Confluence or Notion Basic ($8-15/user/month)
- Required only if the current documentation system causes bottlenecks

Workflow Mapping

- Base: Built-in tools like Microsoft Visio or Draw.io (free)
- When to upgrade: Need real-time collaboration
- Recommended tool: Lucidchart Basic ($7-9/user/month)
- Only needed if the team requires visual process mapping

Project Management

- Base: Trello or Asana Free Tier
- When to upgrade: Need advanced tracking features
- Recommended tool: Basic paid tiers ($10-12/user/month)
- Only needed if: Managing multiple implementation streams

✓ REALITY CHECK

Myth: Technology implementation can succeed without detailed preparation.

Reality: Success comes from systematic readiness planning.

Impact: Poor preparation typically doubles implementation time and costs.

AI Pro Tip: Use our Implementation Readiness Scorecard to identify and address gaps early.

PILLAR 4: TECHNOLOGY & TOOLS – YOUR TECHNICAL INFRASTRUCTURE

In This Section: Learn how to assess your current systems, identify technical needs, and select the right AI solutions. We'll explore systems evaluation, integration requirements, scalability planning, and data capabilities.

Technology & Tools = Existing Systems + Integration Requirements + Scalability Planning + Data Capabilities

CASE STUDY: ENDUP FURNITURE

Just as Tom discovered, their existing systems had untapped potential. *"Everyone told us to start from scratch,"* Tom recalled. *"Instead, we enhanced what we had and saved $30,000 in unnecessary replacements."*

Let me show you how they built their technical foundation across four key areas:

Existing Systems Assessment EndUp started with a thorough evaluation of their current technology:

- Integration readiness
- API availability
- Automation potential
- System compatibility

"Understanding what we already had changed everything," Tom explained. *"We discovered 40% of our systems were already AI-ready."*

Integration Requirements

They developed clear specifications for:

- Data connections
- Workflow integrations
- Security Protocols
- User authentication

Results: Achieved 85% system integration within 60 days.

Scalability Planning

The team established growth requirements for:

- Processing capacity
- Storage needs
- User Expansion

- Feature additions

"Planning for growth from the start saved us from expensive rebuilds later," Tom noted. *"We could scale smoothly as our needs increased."*

Data Capabilities

They focused on critical data elements:

- Data quality standards
- Access protocols
- Storage solutions
- Analysis capabilities

Results After 90 days:

- Integration time was reduced by 60%
- System efficiency improved by 75%
- Data accuracy increased to 99.8%
- Technical support needs decreased by 70%

TECHNOLOGY & TOOLS INFRASTRUCTURE CONSIDERATIONS

(Note: Actual AI tool costs will be covered in Part Two of this chapter)

Time Investment by Business Size:
Small Business (1-10 employees):

- Systems assessment: 2-3 hours
- Integration planning: 2-3 hours
- Data capability review: 2-3 hours
- Technical readiness evaluation: 1-2 hours Total time investment: 7-11 hours in the first month

Medium Business (11-50 employees):

- Systems assessment: 4-6 hours
- Integration planning: 4-6 hours
- Data capability review: 4-6 hours
- Technical readiness evaluation: 2-3 hours Total time investment: 14-21 hours in the first month

Large Business (50+ employees):

- Systems assessment: 8-12 hours
- Integration planning: 8-12 hours
- Data capability review: 8-12 hours
- Technical readiness evaluation: 4-6 hours Total time investment: 28-42 hours in the first month

Potential Technical Upgrades:

Computing Resources

- Base requirement: Modern business-grade computers (past 3-4 years)
- When to upgrade: Systems older than 5 years or below 8GB RAM
- Typical cost, if needed: $600-1,200 per workstation
- Only required if: Current systems cause performance issues

Network Infrastructure

- Base requirement: Standard business internet (50+ Mbps)
- When to upgrade: Multiple users experience slowdowns
- Typical cost, if needed: $100-300/month for faster service
- Only required if: Current speeds impact productivity

Storage Solutions

- Base requirement: Standard business storage (cloud or local)
- When to upgrade: Data needs exceed current capacity
- Typical cost, if needed: $10-50/user/month for additional cloud storage
- Only required if: Current storage limits AI capabilities

Part Two: The AI GROWTH Code

Security Updates

- Base requirement: Standard business security measures
- When to upgrade: Handling sensitive data or compliance requirements
- Typical cost, if needed: $15-30/user/month for enhanced security
- Only required if: Current security doesn't meet AI requirements

> **✓ REALITY CHECK**
>
> **Myth:** You need to replace all existing systems for AI.
>
> **Reality:** Most businesses can leverage and enhance current technology.
>
> **Impact:** Complete system replacement often creates unnecessary disruption.
>
> **AI Pro Tip:** Start by assessing what existing tools can be AI-enhanced.

Important Note: Most modern business systems (less than 3-4 years old) are typically AI-ready. Most businesses we work with require minimal to no infrastructure upgrades before implementing AI solutions.

UNDERSTANDING YOUR AI INFRASTRUCTURE INVESTMENT

(Note: These ranges reflect foundational infrastructure costs and internal resource investments. AI tool costs will be covered In Part Two.)

Small Business (1-10 employees)

Direct Investment Range: $3,000-12,000

Internal Implementation:

- Time Investment: 20-35 hours first month (valued at $2,625)
- Typical Staff Involved: 1-2 key employees
- Leadership Time: 5-10 hours monthly

With AI Pros Support:

- Time Investment: Reduced to 8-15 hours first month (valued at $1,125)
- Typical Staff Involved: 1 key employee
- Leadership Time: 3-5 hours monthly
- Monthly Time-Cost Savings: $1,500
- Implementation Acceleration: 30-45 days faster ROI

Medium Business (11-50 employees)

Direct Investment Range: $10,000-40,000

Internal Implementation:

- Time Investment: 40-70 hours first month (valued at $5,950)
- Typical Staff Involved: 2-4 key employees
- Leadership Time: 10-15 hours monthly

With AI Pros Support:

- Time Investment: Reduced to 15-30 hours first month (valued at $2,550)
- Typical Staff Involved: 1-2 key employees
- Leadership Time: 5-8 hours monthly
- Monthly Time-Cost Savings: $3,400
- Implementation Acceleration: 45-60 days faster ROI

Large Business (50+ employees)

Direct Investment Range: $35,000-100,000

Internal Implementation:

- Time Investment: 80-120 hours first month (valued at $11,400)
- Typical Staff Involved: 4-8 key employees
- Leadership Time: 15-20 hours monthly

With AI Pros Support:

- Time Investment: Reduced to 30-50 hours first month (valued at $4,750)
- Typical Staff Involved: 2-3 key employees
- Leadership Time: 8-12 hours monthly
- Monthly Time-Cost Savings: $6,650
- Implementation Acceleration: 60-90 days faster ROI

Why the Difference?

The AI Pros team's experience with 120+ implementations helps:

- Avoid common pitfalls that waste time
- Provide proven frameworks and templates
- Accelerate decision-making
- Reduce learning curves
- Prevent costly mistakes

NAVIGATING RESOURCE CONSTRAINTS: REAL-WORLD SOLUTIONS

While the investment guidelines provide a framework, reality often throws curveballs. Let me share how different businesses overcame common resource challenges without compromising their AI success.

When the Budget Is Tight

When Green Valley Landscaping faced budget constraints, they got creative. *"Instead of buying new tablets for everyone, we used our existing smartphones for the route optimization app,"* Nina explained. *"That saved us $12,000 in hardware costs."* Their solutions included:

- Using existing hardware creatively
- Starting with free trial versions
- Implementing in phases to spread costs
- Training internal champions instead of hiring consultants
- Result: 30% fuel savings with minimal upfront investment

When Time Is Limited

TRAXX Flooring's team was already working at capacity when they started their AI journey. Their approach:

- 15-minute daily training sessions instead of full-day workshops
- Focus on one process at a time
- Use lunch-and-learn formats for training
- Leverage early adopters to train others
- Result: Achieved implementation goals while maintaining daily operations

When Expertise Is Missing

EndUp Furniture couldn't afford to hire AI specialists, so they built internal capability:

- Identified tech-savvy team members
- Used online learning resources
- Created peer learning groups
- Focused on specific use cases
- Result: Developed in-house expertise at 1/3 the cost of hiring specialists

Myth: You need significant resources to start with AI.

Reality: Creative resource use often beats bigger budgets.

Impact: Resource constraints can actually drive innovation.

AI Pro Tip: Start with what you have and grow systematically.

STRATEGY BY COMPANY SIZE

Small Business (1-10 employees) Limited Resources Solution:

- Use existing hardware
- Focus on one process
- Leverage free tools initially
- Train one champion

Example: A local bakery automated ordering with zero hardware investment

Medium Business (11-50 employees) Balanced Approach:

- Phase investments over 6-12 months
- Combine internal and external expertise
- Prioritize high-impact areas
- Cross-train team members

Example: A manufacturing firm spread $24,000 investment across three quarters

Large Business (50+ employees) Strategic Investment:

- Create a dedicated AI budget
- Build internal capabilities
- Invest in scalable solutions
- Develop training programs

Example: A healthcare provider reduced projected costs by 40% through systematic planning

Making It Work With What You Have

Remember what Mitch from Cornerstone Construction shared: *"We thought we needed everything new. Instead, we learned to maximize what we already had."* Their approach:

1. **Audit Current Resources**

- List all existing technology
- Identify underutilized capabilities
- Map available skill sets
- Result: Found 60% of needed capabilities already in place

1. **Creative Problem-Solving**

- Repurpose existing tools
- Combine resources innovatively
- Share resources across departments
- Result: Reduced new investment needs by half

1. **Staged Implementation**

- Start with zero-cost improvements
- Reinvest early savings
- Build on proven success
- Result: Self-funded 70% of implementation through savings

SUMMARY: BUILDING YOUR COMPLETE AI INFRASTRUCTURE

Through examining each pillar, we've seen how successful companies build their AI foundation:

- People & Skills: Secured Tech Solutions achieved 90% team adoption by focusing on capabilities, training, roles, and leadership
- Budget & Resources: EndUp Furniture reduced implementation costs by 45% through strategic resource planning
- Implementation Readiness: Secured Tech cut response times from 4 hours to 3 minutes with proper preparation
- Technology & Tools: EndUp leveraged existing systems to save $30,000 while improving efficiency by 75%

Key Learnings:

- Each pillar contributes equally (25%) to your overall readiness
- Weakness in any pillar compromises the entire structure
- Success comes from systematic, balanced development
- Strong infrastructure enables faster implementation

Common Infrastructure Pitfalls:

- Overinvesting in technology
- Underinvesting in training
- Rushing implementation
- Ignoring existing capabilities
- Skipping documentation steps

BUILD YOUR AI FOUNDATION: ESSENTIAL RESOURCE PLANNING TOOLS

To help you build the right foundation for your AI initiatives, I've created four critical tools that work together to ensure you have the infrastructure, data, and resources needed for success:

Assessment B - The Resource & Infrastructure Assessment

The comprehensive Resource & Infrastructure Assessment examines your technical readiness, data capabilities, team skills, and available resources. It then provides customized recommendations that have helped organizations prevent costly implementation failures and save tens of thousands in infrastructure rework.

Tool #5 - The Infrastructure Planning Guide

Once you understand your current state, the Infrastructure Planning Guide helps you map out exactly what you need and don't. This practical guide has helped businesses reduce infrastructure costs by 35% while accelerating their implementation timeline by identifying the right investments at the right time.

Tool #6 - The AI Data Requirements Checklist

The AI Data Requirements Checklist ensures you have the data foundation for success. This comprehensive checklist has helped prevent data-related failures that cost organizations $50,000 or more. It's a practical tool that cuts data preparation time in half while ensuring quality and governance.

Tools #7 and #8 AI Talent Management

I've also created two comprehensive online guides focused on AI talent management. The AI Talent Hiring Guide provides a complete framework for recruiting and onboarding AI professionals, helping you avoid costly hiring mistakes and build high-performing teams. The AI Job Specialties Reference Guide, which complements this, breaks down over 40 AI roles and specializations, helping you understand exactly which skills and roles will drive the most value for your specific needs. These guides have helped organizations save hundreds of thousands in hiring mistakes while dramatically reducing the time to value for new AI talent.

Moving Forward: From Infrastructure to Solution Selection

Through Cornerstone's journey, we've seen how building a strong AI Infrastructure creates the foundation for success. But having the right infrastructure is just the beginning. As Mitch noted in our most recent conversation, *"Getting our foundations right was crucial, but then came the real challenge - putting everything into action."*

With your infrastructure foundation mapped out, the next critical decision is to select the right AI tools for your business. In Part Two, we'll explore the six stages of AI-powered solutions, from ready-to-use tools to complex automation systems. Each stage represents a different type of business problem and implementation complexity, but they all build on the infrastructure foundation we just created.

Think of it like a well-equipped workshop. Just as a master craftsman doesn't use a sledgehammer when a precision screwdriver is needed, your Business needs the right AI tools for specific challenges. A small repair shop might need basic power tools for everyday jobs, while a large manufacturing facility requires sophisticated automated systems for complex production. Neither is "better" - they're just right for different needs.

Now that you understand the foundation costs and requirements, you can make informed decisions about what to build on top of it. You'll learn how to match your infrastructure readiness with the right AI capabilities and how to choose solutions that align with both your business goals and implementation capacity.

Remember: The strongest AI implementations don't require the most expensive tools—they need the right tools supported by the right foundation.

RESOURCES (PART TWO) – MATCHING BUSINESS NEEDS TO AI SOLUTIONS

Through helping businesses implement AI over the past three years, I've discovered something crucial: successful AI implementation isn't about choosing the most sophisticated solution - it's about matching the right AI capabilities to your specific business challenges and readiness level. Let me share the framework I use to help business owners understand their options.

I call it "**The 6 Stages of AI-Powered Solutions**." Each stage represents different types of business problems and levels of implementation complexity, from handling individual tasks to automating entire workflows. Think of it like a well-equipped workshop. Just as a master craftsman doesn't use a sledgehammer when a precision screwdriver is needed, or a simple hand saw when the job calls for a precision laser cutter, your business needs the right AI tools for specific challenges. A small repair shop might need basic power tools for everyday jobs, while a large manufacturing facility requires sophisticated automated systems for complex production. Neither is "better" - they're just right for different needs.

The same principle applies to AI solutions: you want tools that match your business challenges, team capabilities, and operational requirements. Using too basic a solution leaves value on the table, while implementing something too complex wastes resources and creates unnecessary risk. You want the solution that best fits your particular business needs and capabilities.

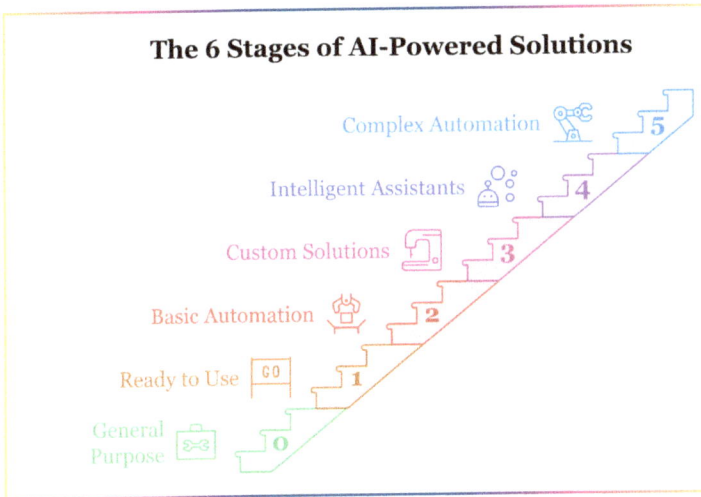

The 6 Stages of AI-Powered Solutions

- Complex Automation — 5
- Intelligent Assistants — 4
- Custom Solutions — 3
- Basic Automation — 2
- Ready to Use — 1
- General Purpose — 0

It's important to note that the field of AI is rapidly evolving, and these stages may overlap or change as new technologies emerge. The goal isn't to race to the top but to find the level that delivers the most value for your business right now.

Let me walk you through each stage and show you how different companies have used them to solve real business problems…

STAGE 0: GENERAL PURPOSE SOLUTIONS

Quick Overview:

Entry-level large language models (LLMs). "Think of Stage 0 solutions like a helpful personal assistant - they can handle individual tasks through conversation, but each request needs to be made separately.

AI LLM Response Generation: From Input to Output

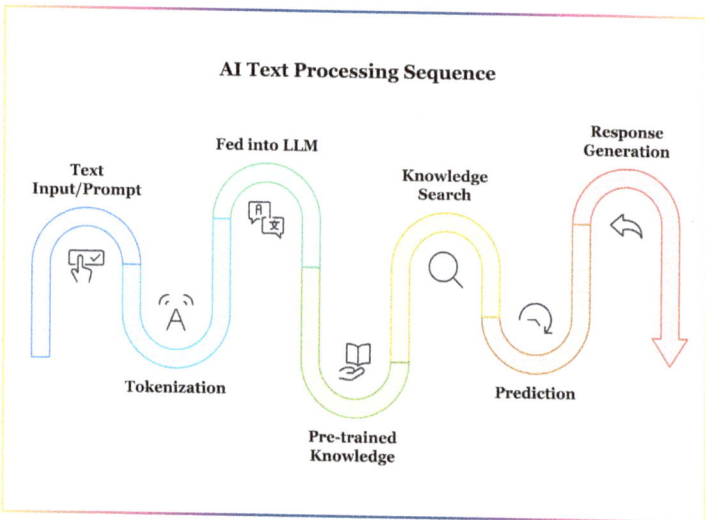

- Text Input/Prompt – You type a request.
- Tokenization – The text is broken into smaller pieces (tokens).
- Fed into LLM – The tokens are processed by the AI model.
- Pre-trained Knowledge – The model uses what it already knows.
- Knowledge Search (if needed) – It looks up real-time info if necessary.
- Prediction – The model predicts the best next words step by step.

- Response Generation – A full response is assembled and sent back to you.

Business Problems Solved:

- Time wasted on repetitive writing tasks
- Hours lost researching basic information
- Creative blocks in content creation
- Slow response times to routine inquiries
- Inconsistent quality in standard communications
- Writing bottlenecks across departments

Popular LLM Models

KEY BENEFITS:

✓ Transform rough drafts into professional content
✓ Distill lengthy documents into key insights
✓ Simplify complex topics into clear explanations

✓ Generate creative ideas and fresh perspectives

Real-World Applications:

➤ Marketing: Creates engaging posts in minutes
➤ Customer Service: Crafts personalized responses quickly
➤ HR: Develops clear, compelling job descriptions
➤ Operations: Captures essential meeting insights efficiently
➤ Training: Produces step-by-step how-to guides
➤ Research: Analyzes market trends and data

Quick Facts:

▪ Typical ROI: 10-20% time savings on routine tasks
▪ Implementation Time: Immediate
▪ Cost Range: $0-1,200/year

Key Limitations:

▪ May provide generic responses rather than business-specific answers
▪ Can occasionally share incorrect information that needs verification
▪ Limited to general knowledge, they were trained on
▪ Cannot access your private business data
▪ No memory between conversations
▪ Cannot run automated processes

Success Snapshot: Stage 0: General Purpose AI

Moriarty's Books invested $240/year in ChatGPT Plus. Their product manager, Sarah, cut the time she spent writing product descriptions from 30 to 8 minutes each, enabling her to list three times more books daily.

ROI: $48,000 annual revenue increase from additional book sales due to faster listing and better descriptions, delivering a 200x return.

6-Month Follow-up:

- Expanded to marketing copy, reducing content creation time from 4 hours to 45 minutes daily, saving an additional 12 hours weekly
- Implemented customer email responses, improving response time by 65%; automated 80% of routine customer inquiries
- Online review ratings increased from 4.2 to 4.7/5

12-Month Impact:

- ROI increased to 300x as the team mastered prompt engineering
- Revenue growth sustained at $72,000 annually, a 32% increase over the baseline
- Monthly book listings increased from 600 to 1,800
- Customer engagement (comments, reviews) up 45%
- Used learnings to inform Stage 1 task automation implementation

Key learning: Starting with a focused writing task shows immediate value and builds confidence in AI capabilities.

Moving from Entry to Ready-to-Use Task-Specific Solutions

While Stage 0 tools excel at individual tasks, many businesses need solutions that can automatically handle specific departmental functions.

Let's look at how Stage 1: Task AI builds on these basic capabilities to deliver specialized value.

Stage 0 to Stage 1 Progression Indicators:

Signs You're Ready to Move from General Purpose to Task AI:

- Your team regularly repeats the same AI prompts
- You need automated rather than manual responses
- Department-specific needs are emerging
- Basic AI use is consistent across team members
- Current solutions require too much manual oversight

STAGE 1: READY-TO-USE SOLUTIONS

Quick Overview:

Pre-built cloud-based smart AI tools that automate specialized tasks within specific departments with little customization needed. Each tool enhances productivity in areas like marketing, sales, design, or customer service.

These are like specialized power tools, each designed for a specific job. Just as you'd use a drill to drill holes and a saw to cut, these tools excel at specific business tasks.

Business Problems Solved

- Critical department tasks taking longer than necessary
- Routine work consuming skilled employees' time
- Inconsistent quality in departmental outputs
- Basic customer inquiries overwhelm support teams
- Manual processes slowing department productivity
- Specialized tasks creating workflow bottlenecks

Common Tools:

Marketing

Social Media

Writing

 LAVENDER

Audio / Video

IIElevenLabs  HeyGen

VEED.IO  loom

 synthesia  Goldcast

 descript CREEK

Key Benefits:

✓ Automate routine department tasks
✓ Maintain consistent quality standards
✓ Deploy ready-to-use customer service automation
✓ Integrate with existing department tools
✓ Handle common customer inquiries with basic chatbots

✓ Support standard business workflows

Real-World Applications:

➤ Marketing: Teams double content production speed
➤ Sales: Personalize outreach at scale
➤ Design: Create professional visuals instantly
➤ Support: Automate common customer questions
➤ Social: Manage multiple channels efficiently
➤ Writing: Maintain style guides automatically

Quick Facts:

- Typical ROI: 20-40% increase in departmental productivity
- Implementation Time: 2-6 weeks
- Cost Range: $600-$3,600/year

Stage 1: Task AI - Key Limitations

- Each tool focuses on one department or function
- Pre-built chatbots limited to general responses
- Cannot customize responses to specific business knowledge
- Requires subscription per tool/department
- Limited cross-department integration
- Need multiple tools for full coverage
- No connection between different tools' data

Success Snapshot: Stage 1: Task AI

EndUp Furniture invested $2,400/year in Lavender AI. The sales team cut email response time from 15 to 3 minutes, enabling them to handle twice the leads.

ROI: $96,000 annual revenue increase from additional closed deals, delivering a 40x return on investment.

6-Month Follow-up:

- Expanded to three departments, maintaining 85% efficiency improvement
- Customer satisfaction increased from 4.2 to 4.8/5
- Lead conversion rate improved from 18% to 25%
- Average order value increased by 15%

12-Month Impact:

- Processing 200+ leads daily with the same staff size (up from 50)
- Customer retention increased from 65% to 82%
- The sales team closed 85% more deals per rep
- Monthly revenue sustained at $12,000 above baseline
- Successfully scaled to Stage 2 automation based on proven ROI

Key learning: Department-specific AI tools show quick wins in daily tasks.

Practical AI Tool Resources (for Stage 1)

As you begin exploring Stage 1 options, here are some valuable directories I recommend to my clients. With new AI tools launching daily, these directories help you stay current and find solutions that match your specific business needs.

TheresAnAIforThat.com

- A comprehensive directory of AI tools
- Searchable by business need
- User reviews and ratings
- Regular updates with new tools

FutureTools.io

- Categorized AI tool listings
- Price comparisons
- Feature breakdowns
- Use case examples

Tools.AI

- Free vs. paid options
- Integration information
- Difficulty ratings
- Implementation guides

RESOURCE TIP: Before investing in any tool, I recommend:

- Reading recent user reviews
- Testing free trials when available
- Checking integration requirements
- Verifying pricing structures
- Evaluating support options

I tell my clients to think of these directories as their AI tool shopping guides. They're constantly updated and can save you hours of research time when looking for the right solution for your specific business needs.

If you need help evaluating or selecting the right tools for your business, my team and I at The AI Pros can guide you through the process. You can book a GROWTH Session <u>HERE</u>.

From Individual Tasks to Process Automation

Stage 1 solutions effectively handle departmental needs, but some businesses require AI that understands their unique industry terminology and processes.

This is where Stage 2: Basic Automation demonstrates its value by streamlining entire workflows.

Stage 1 to Stage 2 Progression Indicators:

Signs You're Ready to Move from Ready-to-use AI to Basic Automation:

- You're connecting multiple tools manually
- Teams follow consistent "if-then" processes
- You need automated workflows rather than individual tasks
- Clear patterns exist in your regular operations
- ROI from Task AI tools has been proven

STAGE 2: BASIC AUTOMATION

Quick Overview:

Rule-based no-code AI automation that transforms repetitive workflows into efficient automated processes.

These solutions follow "if-then" logic: If a specific event (trigger) occurs, certain actions are automatically performed.

Perfect for streamlining predictable workflows, like "if a new support ticket arrives, then route it to the right department" or "if inventory drops below a threshold, then reorder automatically."

Like setting up dominoes - when one falls (trigger), it causes the next to fall (action). Once you set up the pattern, the process runs automatically whenever started.

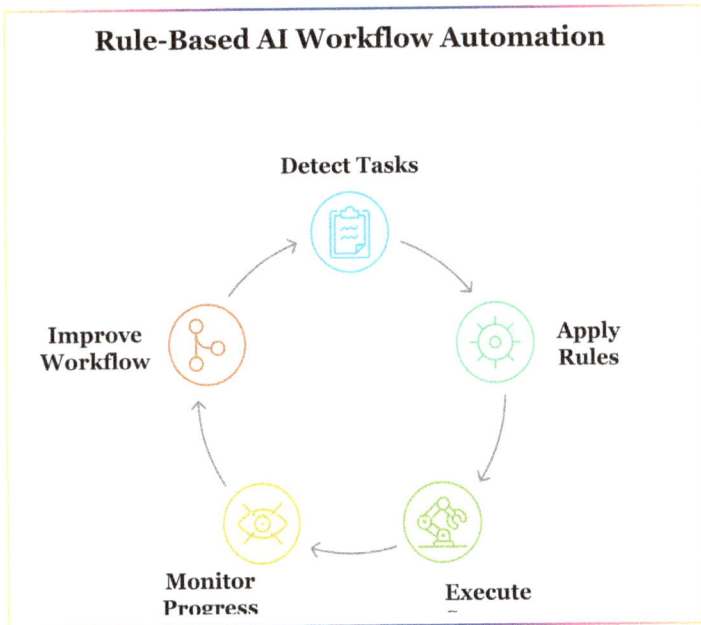

Rule-Based AI Workflow Automation

Detect Tasks

Apply Rules

Execute

Monitor Progress

Improve Workflow

- Detect Task – AI identifies the task based on input.
- Apply Rules – Pre-set rules determine how to handle it.
- Execute Steps – The AI carries out the task automatically.
- Monitor Progress – The system checks if everything runs smoothly.
- Improve Workflow – Adjustments are made to optimize performance.

Business Problems Solved

- High error rates in manual data entry
- Repetitive paperwork consumes staff time
- Inconsistent process execution
- Slow document processing and approvals
- Data trapped in legacy systems
- Time-consuming routine operations

Automation Platforms

activepieces

albaTo

integrately

make
formerly Integromat

mindpal

MindStudio

n8n

Pabbly

Power Automate

relay.app

tray.io

workato

_zapier

Key Benefits:

✓ Significant reduction in processing times
✓ Near-elimination of errors in repetitive processes
✓ Increased efficiency in workflow management

Chapter 8: Resources—Parts One & Two

✓ Improved data accuracy across systems
✓ Consistent application of business rules
✓ Frees employees from mundane tasks

Real-World Applications:

➤ Finance: Process invoices 75% faster
➤ HR: Reduce onboarding time by half
➤ Operations: Eliminate data entry errors
➤ Compliance: Automate routine checks
➤ Purchasing: Streamline order processing
➤ Admin: Complete tasks overnight

Quick Facts:

▪ Typical ROI: 40-60% reduction in processing times
▪ Implementation Time: 2-6 months
▪ Cost Range: $12,000-$60,000/year

Key Limitations:

▪ Works best with structured processes
▪ Requires stable, well-defined workflows
▪ Limited ability to handle exceptions
▪ Needs regular maintenance and updates
▪ May require process redesign
▪ Can't adapt to major process changes
▪ Significant initial investment in setup

Success Snapshot: Stage 2: Basic Automation

TRAXX Flooring invested $15,000 in automation implementation. Processing time dropped from 5 days to 4 hours with zero errors.

ROI: $120,000 annual savings in correction costs plus $200,000 in productivity gains, providing a 21x return.

6-Month Follow-up:

- Automated 85% of routine workflow processes
- Team efficiency increased by 45%
- Error rates reduced from 12% to 0.3%
- Customer response time improved by 60%

12-Month Impact:

- ROI increased to 28x through process optimization
- Processing 3x more orders with the same staff size
- The team reallocated 25 hours weekly to strategic projects
- Successfully expanded to Stage 3 custom solutions

Key learning: Automating structured processes delivers consistent, measurable results.

Advancing to Custom Intelligence

While Stage 2 automation brilliantly handles structured processes, organizations often need solutions that understand their specific business context.

Stage 3: Custom Solutions provides this deeper level of business-specific intelligence.

Stage 2 to Stage 3 Progression Indicators:

Signs You're Ready for Custom Solutions:

- Basic automation needs industry-specific knowledge
- Generic solutions miss important context
- Your processes require unique terminology
- Company-specific compliance needs exist
- The team frequently adjusts standard solutions

STAGE 3: CUSTOM SOLUTIONS

Quick Overview:

AI solutions tailored to your industry knowledge and business processes, trained to understand your specific products, services, and terminology.

These custom GPTs (Generative Pre-trained Transformers) are advanced AI mini-models that understand and generate human-like text. They integrate your proprietary information to provide business-specific responses.

Consider how a veteran employee knows your company's unique terminology and processes. Custom AI is like training a digital employee with your specific business knowledge.

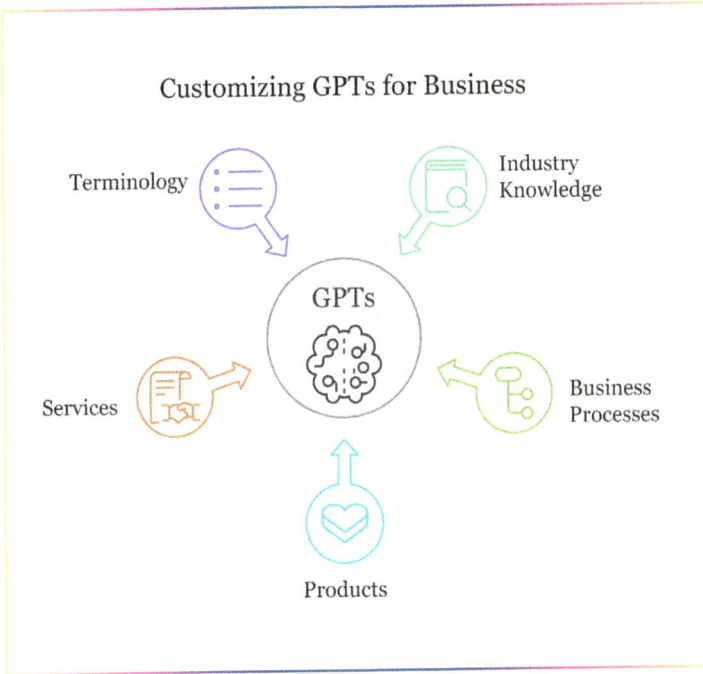

Customizing GPTs for Business

Business Problems Solved

- Generic AI responses missing industry nuance
- Company expertise not being captured or scaled
- Complex industry questions requiring specialist knowledge
- Inconsistent handling of company-specific inquiries
- Industry compliance and regulation challenges
- Need for company-specific automated responses

Common Development Platforms:

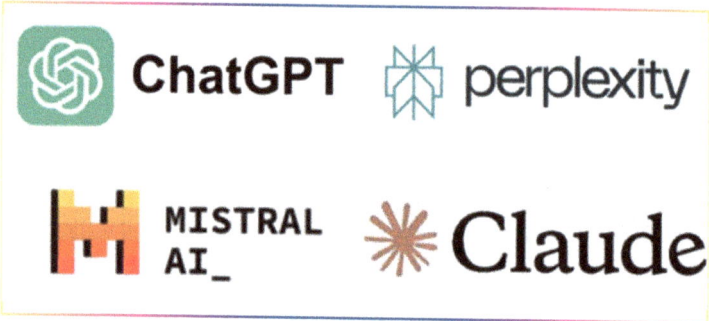

Possible Customizations:

- Industry-specific customer service GPTs
- Custom sales proposal generators
- Specialized document analyzers
- Company-specific knowledge bases
- Product-specific chatbots
- Industry compliance checkers

Key Benefits:

- ✓ Deliver industry-specific responses
- ✓ Apply company policies accurately
- ✓ Handle complex product inquiries
- ✓ Maintain compliance standards
- ✓ Scale internal expertise
- ✓ Provide consistent, accurate information

Real-World Applications:

- ➤ Customer Service: Provides expert-level responses
- ➤ Sales: Generates accurate, customized proposals
- ➤ Compliance: Automates regulatory checks
- ➤ Product Teams: Maintains consistent documentation

- Support: Handles complex technical inquiries
- Training: Scales across departments efficiently

Quick Facts:

- Typical ROI: 30-50% improvement in response accuracy and efficiency
- Implementation Time: 2-8 weeks
- Cost Range: $2,400-$12,000/year

Key Limitations

- Requires initial training with company data
- Needs regular updates to stay current
- More complex to set up than pre-built tools
- Higher initial investment of time and resources
- May not truly understand complex business contexts
- Should include technical expertise for optimal results

Success Snapshot: Stage 3: Custom GPTs

Bon Secours Hospital invested $4,800 in custom GPT development and training. Their HR-specific AI now handles 80% of routine inquiries, saving 90 hours of HR time weekly.

ROI: $120,000 annual savings in staff time while improving response accuracy to 100%. The 25x return enabled HR to focus on strategic initiatives.

6-Month Follow-up:

- Expanded to five departments with 92% accuracy
- Reduced hiring time from 45 to 21 days
- Improved candidate quality scores by 35%
- Training time reduced by 40%

12-Month Impact:

- Staff productivity increased by 65%

- First-year retention improved by 13%
- Recruitment costs are reduced by $65,000 annually
- Successfully integrated with Stage 4 AI agents

Key learning: Training AI on company data transforms it from generic to expert assistant.

FROM CUSTOMIZATION TO AUTONOMOUS OPERATION

Stage 3 Custom Solutions excel at industry-specific tasks, but what if you need AI that can handle complex multi-step processes independently?

This is where Stage 4: AI Agents demonstrate their unique value through autonomous operation.

Stage 3 to Stage 4 Progression Indicators:

Signs You're Ready for AI Agents:

- Need autonomous handling of complex tasks
- Multiple systems require coordination
- Processes span different departments
- Decision-making needs to be automated
- Current solutions require frequent human intervention

STAGE 4: INTELLIGENT ASSISTANTS

Quick Overview:

AI Agents that autonomously handle complex, multi-step tasks. Unlike simpler AI tools that require step-by-step guidance, these agents can independently plan, execute, and adapt entire workflows across multiple platforms and data sources.

Think of AI Agents like a seasoned project manager who can independently coordinate multiple tasks, make decisions, and adapt to changes - all while following your business guidelines.

Business Problems Solved

- Projects requiring 20+ hours of manual coordination
- Research tasks spanning multiple data sources
- Complex analysis requiring days of human effort
- Resource allocation across multiple systems
- Multi-department workflow bottlenecks
- Time-intensive competitive monitoring

Common Development Frameworks:

Example Implementations:

- Market research agents
- Project management assistants
- Data analysis systems
- Customer journey optimizers
- Supply chain monitors
- Competitive analysis tools

Key Benefits:

- ✓ Execute multi-step processes autonomously
- ✓ Coordinate across multiple systems
- ✓ Make decisions based on defined criteria
- ✓ Gather and analyze data independently
- ✓ Adapt workflows based on results
- ✓ Generate comprehensive reports

Real-World Applications:

- ➤ Research: Complete market analysis in hours instead of weeks
- ➤ Projects: Automate resource allocation and tracking
- ➤ Sales: Receive proactive competitive insights
- ➤ Operations: Automatically optimize supply chains
- ➤ Marketing: Capture and analyze customer journeys
- ➤ Finance: Automate complex reporting workflows

Quick Facts:

- Typical ROI: 30-50% increase in productivity for analytical tasks
- Implementation Time: 1-3 months
- Cost Range: $6,000-$24,000/year

Key Limitations:

- Requires significant setup and configuration
- Needs clear boundaries and oversight

- More complex monitoring requirements
- Higher technical expertise needed
- May require custom development
- Careful testing is essential before deployment
- Significant training and maintenance needed

Success Snapshot: Stage 4: AI Agents

Green Valley Landscaping invested $8,000 in the implementation of an AI Agent. The system optimizes routes, monitors weather, and adjusts schedules automatically.

ROI: $180,000 annual savings through 30% fuel reduction and 25% more jobs completed daily, delivering 22x return.

6-Month Follow-up:

- Route efficiency improved by 45%
- Customer satisfaction increased to 4.8/5
- Team overtime reduced by 65%
- Weather-related rescheduling down 80%

12-Month Impact:

- Fuel savings increased to 35%
- Completing 35% more jobs daily
- Revenue per vehicle is up 40%
- Successfully scaled to Stage 5 automation

Key learning: Autonomous AI agents excel when handling multiple variables and complex decisions.

Moving to Enterprise-Wide Intelligence

While Stage 4 AI Agents brilliantly handle complex tasks, some businesses need to consistently automate entire workflows across their organization.

Stage 5: Complex Automation offers this comprehensive capability.

Stage 4 to Stage 5 Progression Indicators:

Signs You're Ready for Complex Automation:

- Enterprise-wide coordination needed
- Multiple AI systems need integration
- Predictive capabilities required
- Real-time adjustments necessary
- Current systems can't scale further

STAGE 5: COMPLEX AUTOMATION

Quick Overview:

Advanced AI systems that integrate with multiple data sources to predict outcomes, automatically adjust processes in real-time, and optimize entire business operations. These systems learn and improve continuously while adapting to changing business conditions.

Similar to how a symphony orchestra coordinates many musicians to create harmonious music, complex automation orchestrates multiple AI systems to work together seamlessly.

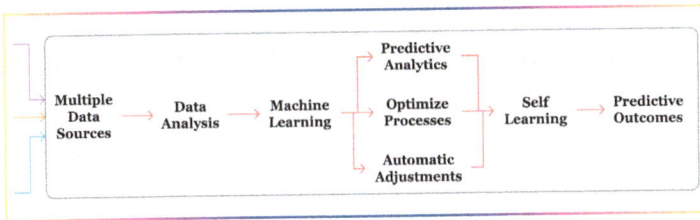

Business Problems Solved:

- Unexpected equipment failures and downtime
- Inefficient supply chain management
- Inaccurate financial forecasting
- Sub-optimal resource allocation
- Inability to predict customer behavior
- Difficulty optimizing complex, interconnected processes

Common Enterprise Automation Platforms:

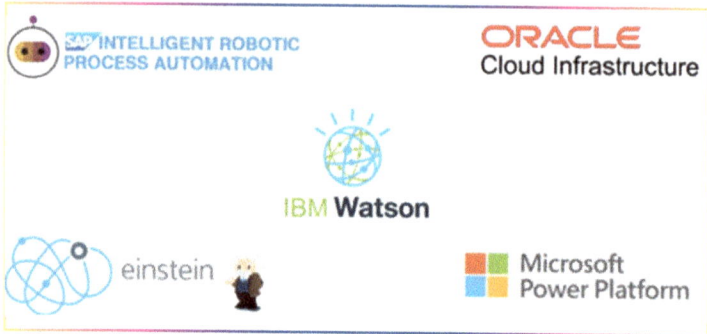

Example Implementations:

- Predictive maintenance systems
- Dynamic supply chain optimization
- Real-time pricing engines
- Automated quality control
- Demand forecasting systems
- Resource optimization platforms

Key Benefits:

- ✓ Dramatic reduction in unexpected downtime
- ✓ Optimized supply chain, leading to reduced costs
- ✓ Highly accurate financial forecasts
- ✓ Increased profitability through optimized pricing
- ✓ Improved customer retention through predictive analytics
- ✓ Significant competitive advantage in rapidly changing markets

Real-World Applications:

- ➤ Manufacturing: Reduces downtime by 70%
- ➤ Supply Chain: Cuts costs by 30%
- ➤ Financial: Improves forecast accuracy by 85%
- ➤ Quality Control: Catches 99% of defects

➤ Operations: Reduces resource waste by 40%

➤ Customer Service: Prevents 60% of issues before they occur

Quick Facts:

- **Typical ROI:** 50-100% improvement in forecast accuracy
- **Implementation Time:** 6-12 months
- **Cost Range:** $25,000-$100,000+/year

Key Limitations:

- High initial investment with delayed ROI
- Potential for AI bias in decision-making processes
- Increased vulnerability to data breaches or cyber attacks
- Over-dependence on AI systems for critical operations
- Requires extensive data infrastructure
- Demands significant implementation time
- Needs specialized expertise and ongoing optimization

Success Snapshot: Stage 5: Complex Automation

Duke's Mayonnaise invested $45,000 in Complex Automation. The system now predicts maintenance needs and automatically optimizes production.

ROI: $500,000 saved in prevented failures, $750,000 gained through efficiency improvements, delivering 27x return in the first year.

6-Month Follow-up:

- Production efficiency increased by 35%
- Quality consistency improved to 97%
- Energy costs reduced by 30%
- Maintenance costs down 45%

12-Month Impact:

- ROI increased to 32x through system optimization
- Downtime reduced by 75%
- Production capacity up 40%
- Quality consistency maintained at 98%
- Successfully integrated across the entire operation

What Do You Need Help With?
The AI Decision Tree

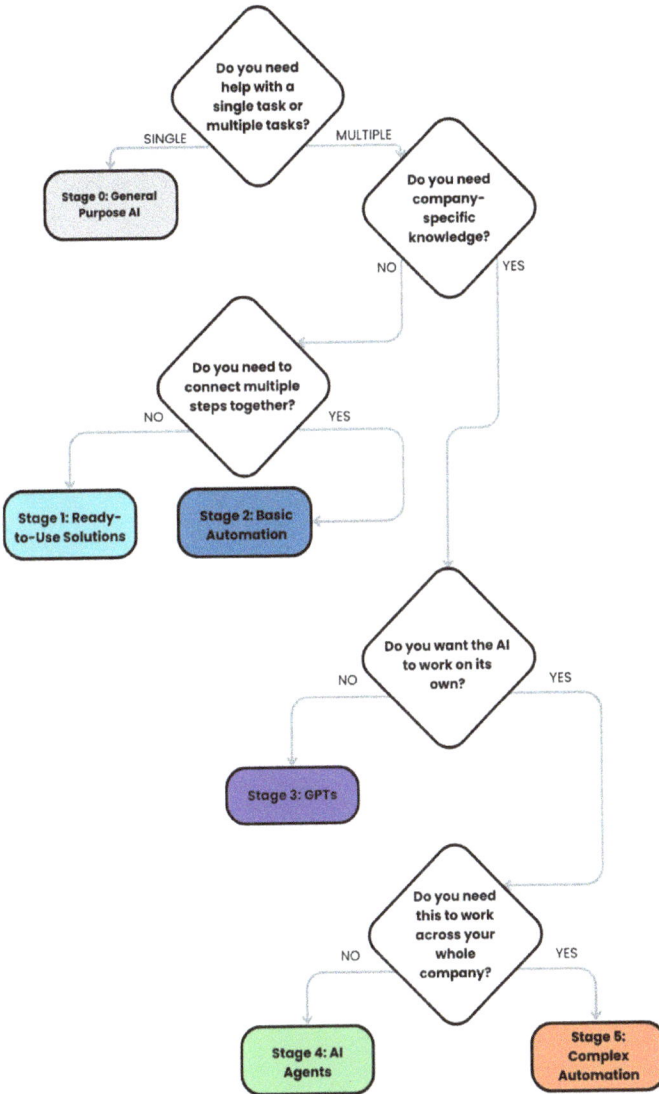

Do you need help with a single task or multiple tasks?

SINGLE → **Stage 0: General Purpose AI**

MULTIPLE → **Do you need company-specific knowledge?**

NO → **Do you need to connect multiple steps together?**

YES → (to "Do you want the AI to work on its own?")

Do you need to connect multiple steps together?
- NO → **Stage 1: Ready-to-Use Solutions**
- YES → **Stage 2: Basic Automation**

Do you want the AI to work on its own?
- NO → **Stage 3: GPTs**
- YES → **Do you need this to work across your whole company?**

Do you need this to work across your whole company?
- NO → **Stage 4: AI Agents**
- YES → **Stage 5: Complex Automation**

Key Takeaways: From Selection to Implementation

Solution Selection Criteria by Stage:

- Stage 0 (Entry AI): Focus on single-task automation
- Stage 1 (Task AI): Department-specific solutions
- Stage 2 (Basic Automation): Process workflow integration
- Stage 3 (Custom Solutions): Industry-specific applications
- Stage 4 (AI Agents): Multi-step autonomous operations
- Stage 5 (Complex Automation): Enterprise-wide intelligence

Implementation Requirements:

Entry Level (Stages 0-1):

- Investment: $0-3,600/year
- Timeline: 1-6 weeks
- Team Skills: Basic digital literacy
- ROI Target: 10-40% efficiency gain

Mid-Level (Stages 2-3):

- Investment: $2,400-12,000/year
- Timeline: 2-8 weeks
- Team Skills: Process expertise
- ROI Target: 30-60% improvement

Enterprise Level (Stages 4-5):

- Investment: $6,000-100,000+/year
- Timeline: 1-12 months
- Team Skills: Advanced technical capability
- ROI Target: 50-100% enhancement

Cost-Benefit Analysis by Industry:

Manufacturing:

- Primary Benefit: Quality control and predictive maintenance
- Typical ROI: 250-350% first year
- Example: Duke's Mayonnaise achieved 27x return

Customer Service:

- Primary Benefit: Response time and satisfaction
- Typical ROI: 200-300% first year
- Example: Secured Tech cut response time by 95%

Financial Services:

- Primary Benefit: Risk management and fraud detection
- Typical ROI: 185-285% first year
- Example: Palmetto Credit Union reduced fraud by 75%

Critical Success Factors:

- Match solutions to business maturity
- Start with proven use cases
- Build on early successes
- Measure ROI consistently
- Scale systematically

Evolution Indicators:

- Reactive to Predictive Operations
- Manual to Automated Processes
- Isolated to Integrated Systems
- Individual to Enterprise Intelligence

MATCH SOLUTIONS TO NEEDS: YOUR AI SELECTION & TRAINING TOOLS

To help you select the right AI solutions and develop your team's capabilities, I've created two essential resources:

Tool #9 - The AI Vendor and Solution Selection Guide

The AI Vendor and Solution Selection Guide simplifies the process of choosing AI vendors and solutions. This step-by-step framework helps you match your business needs to appropriate AI solutions, avoiding costly misalignment. Organizations using this tool have reduced their selection time by 60% while dramatically improving the fit between their needs and chosen solutions.

Tool #10 - The AI Cost-Benefit Analysis and Budgeting Tool

The AI Cost-Benefit Analysis and Budgeting Tool provides a structured framework for evaluating the financial impact of AI initiatives. This tool helps organizations break down the **true costs** of AI implementation—including software, hardware, training, and integration—while also projecting tangible and intangible benefits, such as efficiency gains, revenue growth, and competitive advantages. By using this step-by-step methodology, you can accurately assess ROI, compare multiple AI investment scenarios, and build a realistic AI budget that aligns with your strategic objectives.

MOVING FORWARD: FROM FOUNDATION AND SOLUTIONS TO IMPLEMENTATION

Through Chapter 8, we've established two critical elements: first, how to build your AI Infrastructure across four essential pillars, and second, how to select the right AI solutions for your specific business needs. Now it's time to put both pieces into action.

This brings us to the 'O' in our GROWTH Code - Operationalize. In Chapter 9, we'll explore exactly how to activate your infrastructure and implement your chosen solutions in daily operations. You'll learn how companies like Cornerstone Construction turned their strong foundations into working systems by selecting and implementing the right AI tools and how Green Valley Landscaping transformed their route optimization from a pilot project into a company-wide operation.

Remember: A solid infrastructure and the right AI solutions create potential, but operationalization turns that potential into results. Now, let's transform your AI foundation and tools into business reality.

CITATIONS:

[3] https://www.leewayhertz.com/ai-in-business-process-automation/

[4] https://contentsquare.com/blog/ai-tools-for-advanced-analytics/

[5] https://productschool.com/blog/artificial-intelligence/ai-business-use-cases

[6] https://www.pecan.ai/blog/deep-dive-advanced-predictive-analytics/

[7] https://cloud.google.com/transform/101-real-world-generative-ai-use-cases-from-industry-leaders?hl=en

.

Part Two: The AI GROWTH Code

CHAPTER 9

OPERATIONALIZE–
SEAMLESSLY INTEGRATE AI
WITHIN CURRENT SYSTEMS

The AI
Growth
Code™

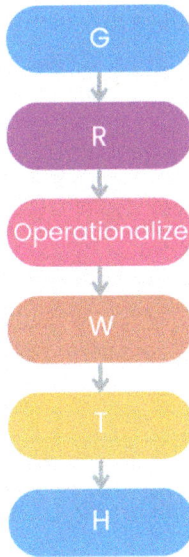

When Tom from EndUp Furniture reviewed his SMART goals, he had a critical question: *"Rich, we've defined our objectives clearly, but which should we tackle first?"*

This brings us to a crucial step in the GROWTH Code - Operationalize.

The key to successful implementation is prioritizing your goals effectively.

Transforming SMART Goals into Action

Let's start by reviewing Tom's SMART goals:

(From Chapter 7)

Implement an AI payment prediction and automation system within 90 days to reduce the percentage of accounts over 30 days late from 35% to 15%. This will improve monthly cash flow by $42,000 while maintaining customer satisfaction above 4.5/5 and reducing collection team time by 3 hours daily.

THE IMPACT/EFFORT MATRIX: YOUR PRIORITIZATION TOOL

To help businesses prioritize their AI initiatives effectively, I've developed the Impact/Effort Matrix. This simple but powerful tool enables you to evaluate and categorize your goals based on two critical factors:

Impact: The potential business value and ROI

Effort: The resources, time, and complexity required

The matrix divides projects into four categories:

- **Quick Wins:** High Impact/Low Effort
- **Strategic Projects:** High Impact/High Effort
- **Fill-ins:** Low Impact/Low Effort
- **Long-term Projects:** Low Impact/High Effort

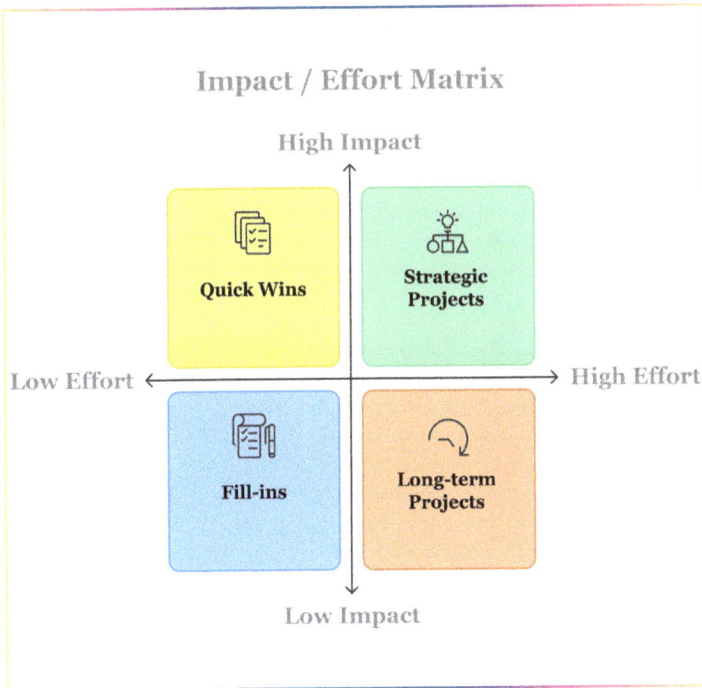

Turning SMART Goals into Prioritized Action

When Tom reviewed his SMART goals, he needed to prioritize them systematically. Let me show you how we used the Impact/Effort Matrix to evaluate his first goal:

Impact Assessment:

- Does it affect revenue or cash flow? *Yes, improves cash flow by $42,000 monthly*
- Can it be implemented quickly? *Yes, using existing A/R data*
- Will it show immediate results? *Yes, measurable within 30 days*
- Does it solve a pressing business problem? *Yes, late payments affecting operations*

Effort Evaluation:

- Do we have the necessary data? *Yes, in the current A/R system*
- Are required skills available? *Yes, the team already manages collections*
- Is the technology accessible? *Yes, basic AI tools can handle this*
- Can we implement it with minimal disruption? *Yes, enhances an existing process*

Result: This goal rates as High Impact (immediate cash flow improvement) with Low Effort (uses existing data and systems), making it an ideal Quick Win.

"This evaluation process was eye-opening," Tom shared. *"Instead of trying to tackle everything at once, we could see exactly where to start."*

Industry-Specific Examples

Let me show you how this matrix applies across different industries:

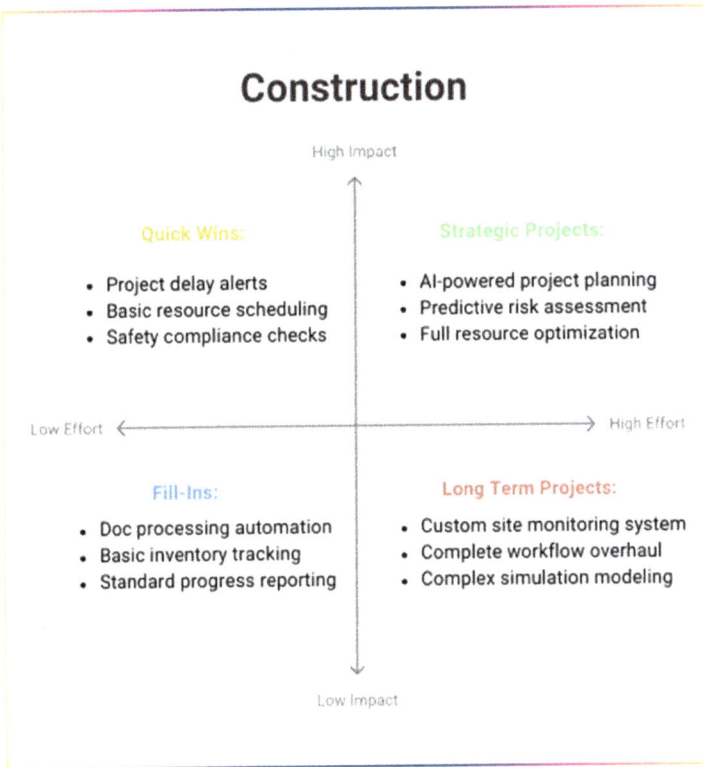

Chapter 9: Operationalize- Seamlessly Integrate AI Within Current Systems

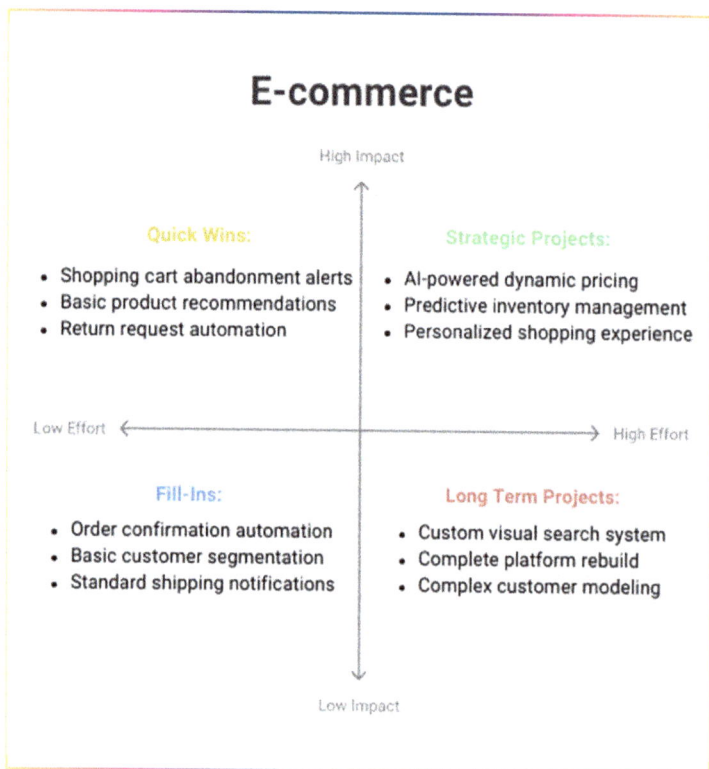

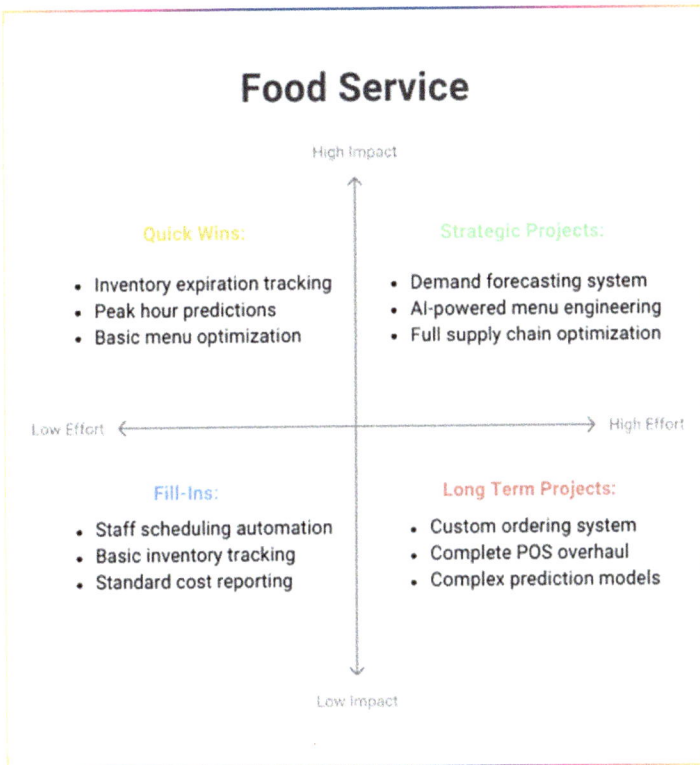

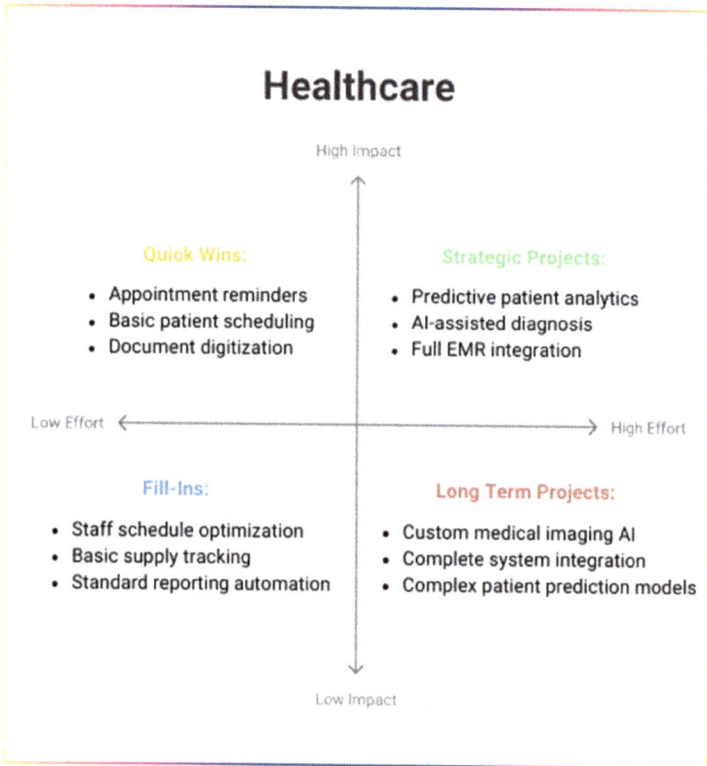

Healthcare

High Impact

Quick Wins:
- Appointment reminders
- Basic patient scheduling
- Document digitization

Strategic Projects:
- Predictive patient analytics
- AI-assisted diagnosis
- Full EMR integration

Low Effort ⟵⟶ High Effort

Fill-Ins:
- Staff schedule optimization
- Basic supply tracking
- Standard reporting automation

Long Term Projects:
- Custom medical imaging AI
- Complete system integration
- Complex patient prediction models

Low Impact

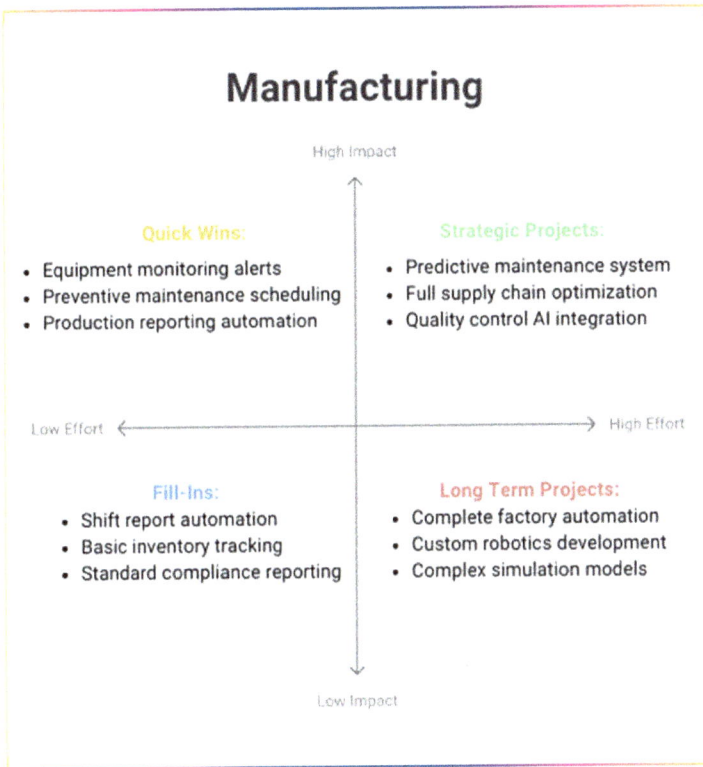

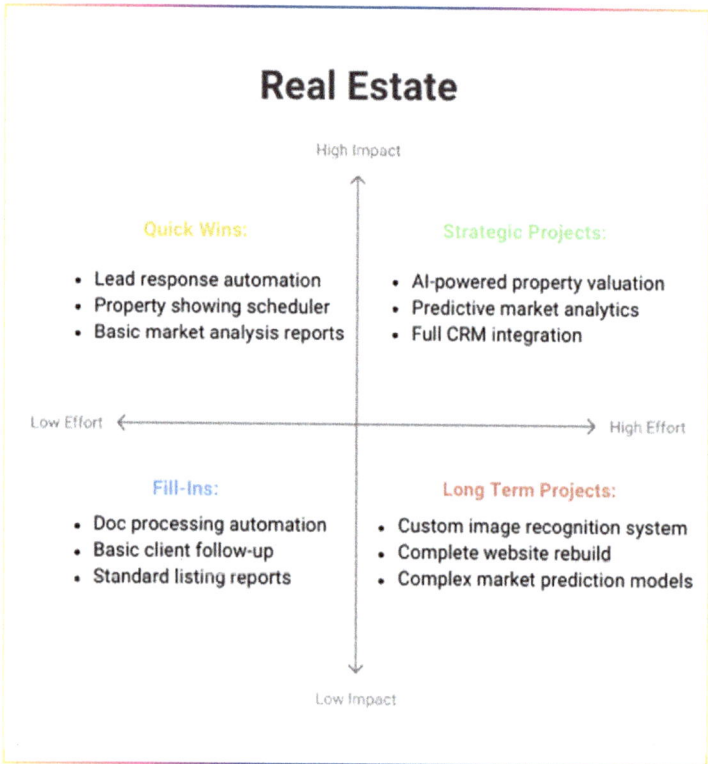

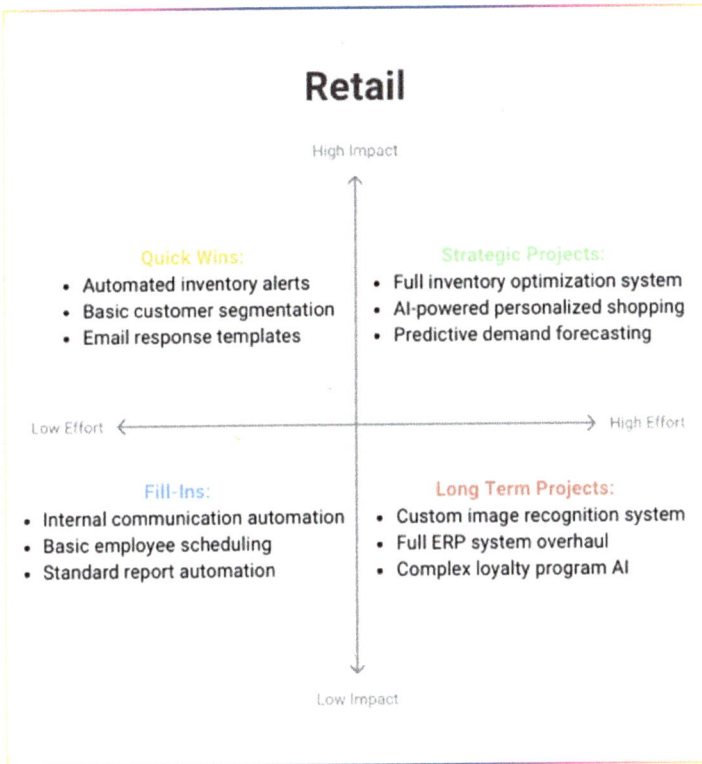

Retail

High Impact

Quick Wins:
- Automated inventory alerts
- Basic customer segmentation
- Email response templates

Strategic Projects:
- Full inventory optimization system
- AI-powered personalized shopping
- Predictive demand forecasting

Low Effort ← → High Effort

Fill-Ins:
- Internal communication automation
- Basic employee scheduling
- Standard report automation

Long Term Projects:
- Custom image recognition system
- Full ERP system overhaul
- Complex loyalty program AI

Low Impact

Chapter 9: Operationalize- Seamlessly Integrate AI Within Current Systems

PLACING YOUR GOALS IN THE MATRIX

To place your SMART goals effectively, evaluate each one through two lenses:

Impact Assessment:

- Financial impact
- Operational efficiency
- Customer relationships
- Team Productivity

Effort Evaluation:

- Implementation complexity
- Resource requirements
- Team training needs
- Integration demands

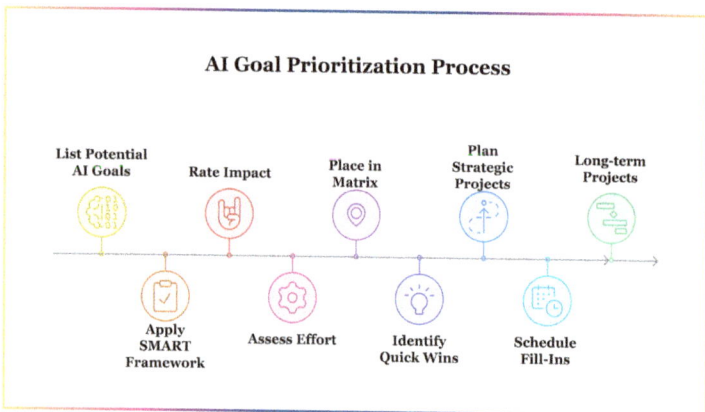

AI Goal Prioritization Process

List Potential AI Goals — Rate Impact — Place in Matrix — Plan Strategic Projects — Long-term Projects

Apply SMART Framework — Assess Effort — Identify Quick Wins — Schedule Fill-Ins

PLACING ENDUP'S GOALS IN THE MATRIX

Using this evaluation framework, here's how we placed Tom's goals:

Quick Wins (High Impact/Low Effort):

- Basic payment reminder automation
- High Impact: Directly improves cash flow
- Low Effort: Uses existing systems
- Quick ROI: Expected within 30-45 days

Strategic Projects (High Impact/High Effort):

- Full AR system overhaul
- High Impact: Complete process transformation
- High Effort: Major system integration required
- Longer ROI: 6-12 months

Fill-ins (Low Impact/Low Effort):

- Basic payment tracking automation
- Lower Impact: Internal efficiency improvement
- Low Effort: Simple process change
- Supporting Role: Enhances main initiatives

Long-term Projects (Low Impact/High Effort):

- Custom financial analytics platform
- Future Impact: Long-term potential
- High Effort: Significant development needed
- Extended Timeline: 12+ months

Chapter 9: Operationalize- Seamlessly Integrate AI Within Current Systems

THE PRIORITY SEQUENCE

Through implementing AI in over 120 businesses, I've discovered a clear sequence for success:

Start with Quick Wins

- Build confidence through early success
- Generate immediate ROI
- Gain team buy-in
- Create Momentum

Move to Strategic Projects

- Build on proven success
- Apply lessons learned
- Tackle larger transformations
- Drive major improvements

Add Fill-ins

- Support primary initiatives
- Maintain momentum
- Complete foundation work
- Enhance overall efficiency

Consider Long-term Projects

- Plan for future growth
- Build comprehensive solutions
- Transform operations
- Create lasting change

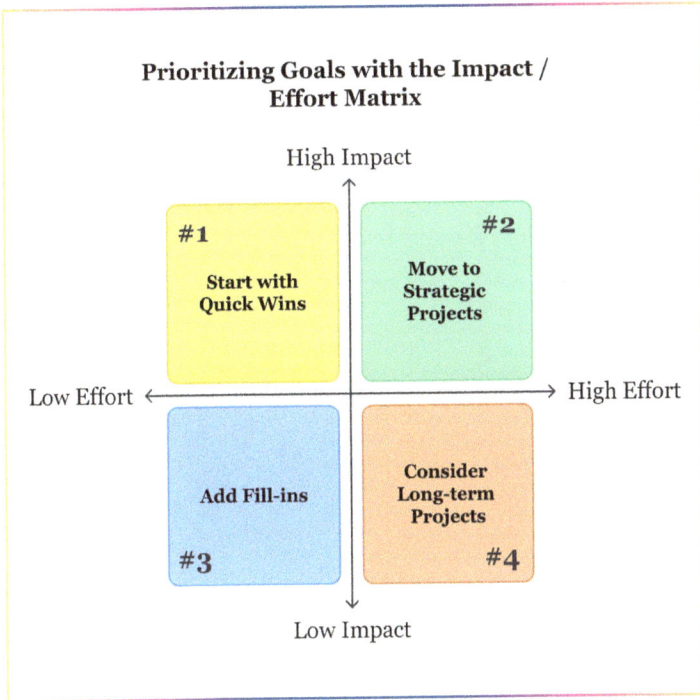

Prioritizing Goals with the Impact / Effort Matrix

High Impact

#1 Start with Quick Wins	#2 Move to Strategic Projects

Low Effort ← → High Effort

Add Fill-ins #3	Consider Long-term Projects #4

Low Impact

Let me share specific Quick Win success stories that demonstrate this prioritization approach in action:

QUICK WINS IN ACTION: REAL IMPLEMENTATION STORIES

EndUp Furniture's First Victory

- SMART Goal: Reduce accounts over 30 days late from 35% to 15%
- Quick Win Selected: Basic payment reminder automation
- Implementation Timeline: 4 weeks
- Initial Investment: $2,400
- Results:

- Late payments reduced by 25% in the first 30 days
- The team saved 2 hours daily on collections
- ROI achieved in 28 days
- Provided foundation for larger AR transformation

"Starting with automated reminders proved our concept quickly," Tom shared. *"That early success made getting buy-in for bigger projects easier."*

Green Valley Landscaping's Route to Success

- SMART Goal: Reduce fuel costs by 30% within 90 days
- Quick Win Selected: Basic route optimization for two crews
- Implementation Timeline: 2 weeks
- Initial Investment: $1,800

Results:

- 30% fuel savings for test crews
- 90 minutes saved daily in planning
- ROI achieved in 15 days
- Created template for full fleet roll-out

"That first success with just two crews," Nina explained, *"showed everyone exactly what was possible."*

Secured Tech Solutions' Support Transformation

- SMART Goal: Reduce response time from 4 hours to under 30 minutes
- Quick Win Selected: Basic ticket routing automation
- Implementation Timeline: 3 weeks
- Initial Investment: $2,400

Results:

- Response time dropped to 15 minutes
- 40% of tickets automatically categorized

- ROI achieved in 21 days
- Built foundation for complete support system overhaul

Let me show you how to identify the right Quick Win for your industry.

Identifying Quick Wins By Industry

Let me show you how different industries can identify and implement their ideal Quick Wins. These examples come directly from successful implementations across my client base.

Construction Industry Quick Wins

Cornerstone Construction found its first success with project delay alerts:

- Initial Challenge: Projects running 20% over schedule
- Quick Win: AI-powered project delay prediction
- Implementation: 3 weeks, $2,400 investment

Results:

- 35% reduction in delays
- 25% improvement in resource allocation
- ROI achieved in 32 days

E-COMMERCE QUICK WINS

EndUp Furniture's online division tackled cart abandonment:

- Initial Challenge: 75% cart abandonment rate
- Quick Win: Automated abandonment alerts and follow-up
- Implementation: 2 weeks, $1,800 investment

Results:

- The recovery rate improved 40%
- $28,000 monthly revenue increase
- ROI achieved in 19 days

FINANCIAL SERVICES QUICK WINS

Palmetto Credit Union started with transaction anomaly alerts:

- Initial Challenge: Growing fraud losses
- Quick Win: Basic fraud detection system
- Implementation: 4 weeks, $3,600 investment

Results:

- 65% reduction in fraud losses
- $45,000 saved in the first quarter
- ROI achieved in 24 days

FOOD SERVICE QUICK WINS

Duke's Mayonnaise began with inventory expiration tracking:

- Initial Challenge: $4,000 monthly inventory waste
- Quick Win: Basic expiration prediction system
- Implementation: 3 weeks, $2,400 investment

Results:

- Waste reduced by 70%
- $2,800 monthly savings
- ROI achieved in 26 days

HEALTHCARE QUICK WINS

Bon Secours Hospital focused on appointment reminders:

- Initial Challenge: 22% no-show rate
- Quick Win: AI-powered reminder system
- Implementation: 2 weeks, $1,800 investment

Results:

- The no-show rate dropped to 8%
- $38,000 monthly revenue increase
- ROI achieved in 15 days

MANUFACTURING QUICK WINS

TRAXX Flooring implemented equipment monitoring alerts:

- Initial Challenge: Unexpected equipment failures
- Quick Win: Basic predictive monitoring
- Implementation: 4 weeks, $3,600 investment

Results:

- Downtime reduced by 45%
- Maintenance costs down 30%
- ROI achieved in 35 days

REAL ESTATE QUICK WINS

Mason & Magnolia started with lead response automation:

- Initial Challenge: 4-hour average response time
- Quick Win: Automated initial responses
- Implementation: 2 weeks, $1,800 investment

Results:

- Response time under 5 minutes
- Lead conversion up 35%
- ROI achieved in 21 days

RETAIL QUICK WINS

Plush Leg Warmers implemented inventory alerts:

- Initial Challenge: Stock-outs and overstock issues
- Quick Win: Basic inventory prediction
- Implementation: 3 weeks, $2,400 investment

Results:

- Stock-outs reduced by 60%
- Carrying costs down 25%
- ROI achieved in 28 days

KEY SUCCESS PATTERNS

Across all these successful Quick Win implementations, certain patterns emerged:

Clear Focus

- Single, specific problem
- Measurable outcomes
- Limited scope
- Defined success criteria

Rapid Implementation

- 2-4 week timeline
- Minimal disruption
- Basic training needs
- Fast feedback loop

Visible Results

- Clear metrics
- Easy measurement

- Team-level impact
- Customer benefits

Low Risk

- Small initial Investment
- Limited process changes
- Easy rollback option
- Manageable scope

MOVING BEYOND QUICK WINS

I've identified clear signals that indicate your Quick Win implementation is ready to grow:

Success Indicators:

Quantitative Metrics

- ROI achieved within the planned time frame
- Stable or improving performance
- Consistent data quality
- Measurable business impact

Team Engagement

- High user adoption rates (>80%)
- Confident system usage
- Proactive improvement suggestions
- Positive feedback from stakeholders

System Stability

- Reliable daily operation
- Minimal technical issues
- Strong data accuracy
- Smooth workflow integration

"We knew we were ready for more when our team started asking what else we could automate," Tom from EndUp Furniture shared. *"The Quick Win had shifted their thinking from 'Will this work?' to 'What's next?'"*

Common Quick Win Implementation Challenges

Let me share the most common challenges I see during Quick Win implementation and their solutions:

Data Quality Issues

- Challenge: Inconsistent or incomplete data
- Solution: Start with basic data cleanup in one area
- Example: Secured Tech focused on standardizing ticket categories before automation

Team Resistance

- Challenge: Initial skepticism about AI
- Solution: Involve skeptics in the implementation process
- Example: Nina at Green Valley made her most skeptical driver the first route optimization tester

Scope Creep

- Challenge: Trying to add features during implementation
- Solution: Maintain a strict Quick Win focus
- Example: EndUp Furniture kept their first automation focused solely on payment reminders

Measurement Confusion

- Challenge: Unclear success metrics
- Solution: Establish specific KPIs before starting
- Example: Bon Secours defined exact targets for appointment no-show reduction

LEARNING FROM FAILED IMPLEMENTATIONS: RECOVERY STORIES

Let me share some costly lessons that shaped my approach to AI implementation. These aren't just failures—they're blueprints for success.

The All-In Disaster A manufacturing firm invested $85,000 in AI tools and attempted to transform their entire production process at once. *"We thought bigger was better,"* their Operations Director admitted. Within three months:

- Team overwhelmed by changes
- Production delays increased by 25%
- Quality control suffered
- Employee morale plummeted

The Recovery: We paused everything and restarted with one production line. By focusing on quality control in that single line, they improved by 35% in 60 days. This success naturally drew the attention of other departments.

The Skip-the-Training Shortcut An accounting firm rushed to implement AI document processing without proper training. The results were predictable:

- 65% error rate in document classification
- Staff reverted to manual processing
- $28,000 wasted on unused licenses
- Client complaints increased

The Recovery: Implemented a "Train First, Deploy Second" approach. Started with two-hour training sessions for core users. Within 45 days, they achieved 95% accuracy and expanded successfully.

The Copy-Paste Problem A retail chain tried to replicate another company's successful AI implementation exactly:

- Wrong tool for their specific needs
- Processes didn't match their workflow
- Team resistance due to poor fit
- $42,000 spent with no results

The Recovery: Started fresh with a proper assessment of their unique needs. Chose appropriate tools and achieved positive ROI within 90 days.

The Data Disaster A distribution company implemented AI without cleaning their data:

- AI made wrong predictions
- The team lost faith in the system
- Customer orders mishandled
- Three months of chaos

The Recovery: Spent two weeks cleaning historical data. Created data quality protocols. Now maintaining 99.8% accuracy in order processing.

Key Recovery Patterns

These failures revealed consistent recovery patterns:

1. Stop and Reset

- Pause implementation
- Assess current state
- Gather team feedback
- Create recovery plan

2. Start Small

- Choose one process
- Build team confidence

- Document everything
- Celebrate small wins

   3. Scale Smart

- Use proven success
- Train thoroughly
- Monitor closely
- Adjust as needed

✓ REALITY CHECK

Myth: Failed implementations mean AI won't work for your business.

Reality: Most failures become successes with the right recovery approach.

Impact: Learning from failures often leads to stronger implementations.

AI Pro Tip: Document what went wrong to prevent repeat mistakes.

QUICK WIN TROUBLESHOOTING GUIDE

Common Issue -> Quick Solution

- Data Quality Problems -> Start with 2 weeks of manual data cleaning before AI implementation
- Team Resistance -> Begin with your most enthusiastic team member as system champion
- System Integration -> Use existing APIs before attempting custom development
- Performance Issues -> Focus on perfecting one process before scaling

Chapter 9: Operationalize- Seamlessly Integrate AI Within Current Systems

Real Example: When Tony faced initial data quality issues with customer records, he spent two weeks standardizing his booking information. *"Those two weeks of preparation saved us months of potential problems,"* he noted.

Creating Your Quick Win Action Plan

Before implementing your Quick Win, use this proven framework:

Pre-Implementation Checklist

- Confirm SMART goal alignment
- Verify high-impact/low-effort classification
- Document current baseline metrics
- Define specific success criteria
- Identify key stakeholders
- Assess resource requirements
- Create timeline milestones
- Establish communication plan

IMPLEMENTATION SCHEDULE

Week 1:

- Team Preparation
- Basic training
- System setup
- Initial testing

Week 2-3:

- Pilot deployment
- User feedback collection
- Process refinement
- Performance monitoring

Week 4:

- Full deployment
- Performance measurement
- Success verification
- Documentation completion

Success Metrics Framework

Track these key indicators:

- ROI achievement
- User adoption rates
- Process efficiency gains
- Quality improvements
- Customer impact
- Team satisfaction

Standardizing Your AI Quick Win: The SOP Framework

One crucial element for Quick Win's success is having clear, documented procedures. Here's an example of an AI-focused SOP template.

AI OPERATIONS SOP TEMPLATE EXAMPLE

Download a copy at aigrowthcode.com

Document Control

- SOP Title: AI-Enhanced Process Management
- Version: 1.0
- AI System Owner: [Role]
- Process Owner: [Department Lead]

Purpose & Scope

- Define AI system interaction protocols
- Establish human-AI collaboration workflow
- Set performance monitoring standards
- Define exception-handling procedures

Roles & Responsibilities

- AI System Monitor: Oversees AI performance
- Process Users: Daily AI system interaction
- Quality Control: Reviews AI decisions
- Technical Support: Manages AI maintenance

Daily AI Operations

System Health Check

- AI system status verification
- Data quality assessment
- Performance metrics review
- Exception queue check

Core AI Process Steps

- AI processing parameters
- Human oversight triggers
- Quality control checkpoints
- Feedback loop protocols

AI Exception Handling

- Undefined scenario protocols
- Manual review triggers
- Escalation procedures
- Learning update process

AI Performance Review

- Accuracy metrics analysis
- Learning curve assessment
- System optimization needs
- Knowledge base updates

AI Performance Monitoring

- System accuracy tracking
- Processing speed metrics
- Learning rate assessment
- Error pattern analysis

ADAPTING AI SOPS FOR YOUR BUSINESS SIZE

Through helping businesses implement AI across different scales, I've learned that effective SOPs must match an organization's size and complexity. Let me show you how other companies have successfully adapted their AI SOPs.

Small Business Adaptation (1-10 employees)

Plush Leg Warmers' AI Pricing SOP:

Core Elements:

Daily AI System Check (15 minutes)

- Verify pricing recommendations
- Check competitor data feeds
- Review exception flags

Twice-Daily Review Points

- Morning Market Analysis Review
- Afternoon price adjustment approval

Simple Exception Protocol

- Clear override criteria
- Basic escalation path
- Documentation requirements

"We kept it focused and practical," Katie explained. *"Every step had a clear purpose, and everyone knew exactly what to do."*

Mid-Size Implementation (11-50 employees)

Secured Tech Solutions' Support System SOP:

Key Components:

Shift-Based AI Monitoring

- System health checks every 4 hours
- Cross-team handover protocols
- Regular accuracy assessments

Tiered Response Framework

- AI-first response review
- Human escalation triggers
- Quality control checkpoints

Department Integration

- Support team protocols
- Technical team responsibilities
- Management oversight requirements

"The key was finding the right balance between automation and human oversight," Joshua shared. *"Our SOP made those boundaries clear."*

Enterprise Scale (50+ employees)

Duke's Mayonnaise Production AI SOP:

Comprehensive Framework:

Multi-Shift Operations

- 24/7 AI system monitoring
- Shift handover procedures
- Cross-line coordination
- Emergency response protocols

Production Integration

- AI quality control checks
- Recipe optimization protocols
- Predictive maintenance responses
- Inventory management rules

Compliance & Documentation

- Regulatory requirement checks
- Quality assurance protocols
- Audit trail maintenance
- Training verification

"With multiple production lines running 24/7," Mike explained, *"our SOP had to be comprehensive but still user-friendly. It became our operational backbone."*

KEY SOP SUCCESS PATTERNS

Regardless of size, successful AI SOPs share these characteristics:

Clear AI-Human Interaction Points

- When AI operates independently
- When a human review is required
- How to handle exceptions
- Who makes the final decisions

Simple Documentation Requirements

- Essential data capture
- Decision logging
- Exception tracking
- Performance monitoring

Defined Learning Loop

- How to flag AI errors
- Process for system updates
- Performance improvement tracking
- Knowledge base maintenance

Scalable Structure

- Core processes that can grow
- Modular components
- Clear upgrade paths
- Flexible integration points

"The best SOPs grow with you," Katie from Plush Leg Warmers noted. *"We started simple but built-in room to expand."*

INDUSTRY-SPECIFIC AI SOP EXAMPLES

Based on successful implementations across different sectors, here's how various industries adapted their AI SOPs to their unique needs:

Construction Industry

Cornerstone Construction's Project Delay Alert SOP:

<u>Daily AI Monitoring</u>

- Morning Project Scan (7:00 AM)
- AI system health check
- Delay prediction review
- Resource allocation flags
- Weather impact assessment

<u>Alert Response Protocol</u>

- Risk Level Categories
- Low: 24-hour response window
- Medium: 4-hour response
- High: Immediate action required

<u>System Learning</u>

- Daily Feedback Loop
- Prediction accuracy tracking
- False alert documentation
- Pattern Recognition updates
- Resource impact analysis

E-COMMERCE

EndUp Furniture's Cart Abandonment SOP:

Real-Time Monitoring

AI Alert Thresholds

- Abandonment pattern detection
- Customer behavior triggers
- Response timing rules
- Follow-up sequences

Recovery Process

- Automated Response Tiers
- Immediate chat engagement
- 1-hour email window
- 24-hour remarketing
- Weekly trend analysis

FINANCIAL SERVICES

Palmetto Credit Union's Fraud Detection SOP:

System Verification

- Hourly Checks
- AI alert review
- Pattern analysis
- Risk score validation
- Regulatory compliance

Response Protocols

- Risk-Based Actions
- Low: System flag

- Medium: Human review
- High: Immediate freeze
- Critical: Law enforcement contact

HEALTHCARE

Bon Secours Hospital's Appointment Management SOP:

<u>Daily Operations</u>

- Schedule Optimization
- AI capacity analysis
- No-show prediction
- Resource allocation
- Emergency slot management

<u>Patient Communication</u>

- Automated Workflows
- Reminder sequences
- Confirmation tracking
- Rescheduling prompts
- Follow-up protocols

MANUFACTURING

TRAXX Flooring's Equipment Monitoring SOP:

Continuous Monitoring

- AI System Checks
- Performance metrics
- Anomaly detection
- Maintenance triggers
- Quality indicators

Response Framework
- Alert Levels
- Warning: Schedule check
- Alert: Immediate inspection
- Critical: Emergency shutdown
- Trend: Weekly review

BEST PRACTICES ACROSS INDUSTRIES

Through these implementations, we've identified universal SOP elements that drive success:

Clear Trigger Points
- When AI acts independently
- When a human review is needed
- Emergency override protocols
- Learning update triggers

Response Time Standards
- Real-time actions
- Near-term responses
- Scheduled reviews
- Learning cycles

Documentation Requirements
- System decisions
- Human interventions
- Exception handling
- Performance metrics

Learning Integration
- Success tracking
- Error documentation

- Pattern updates
- System optimization

ALTERNATIVE IMPLEMENTATION PATHS

When standard Quick Wins don't fit your situation:

- Zero Budget Start: Begin with free AI tools for proof of concept
- Limited Data: Start with manual data collection for 30 days
- High Regulation: Focus on non-regulated processes first
- Remote Teams: Implement cloud-based solutions

Each path still follows our core principle: start small, prove value, then expand.

Operationalize with Confidence: Your Implementation Toolkit

To help you move from planning to successful execution, I've developed three essential tools that guide you through every aspect of AI operationalization:

Assessment C - Implementation Readiness Assessment

Combined with our Implementation Readiness Assessment, which evaluates your operational preparedness across processes, people, and systems, you'll know exactly where to focus your efforts first.

Tool #11 - Impact/Effort Matrix Tool

Start with the Impact/Effort Matrix Tool - a simple but powerful framework for prioritizing your AI initiatives. This four-quadrant approach helps you identify quick wins that can deliver rapid ROI while building momentum for larger projects.

Tool #12 - AI Implementation SOP Template

The AI Implementation SOP Template helps you establish clear procedures and plan high-impact initiatives that deliver results quickly. Organizations using this template have cut training time nearly in half while significantly improving success rates.

LOOKING AHEAD

In our next chapter, we'll explore how to "Widen" your AI implementation, building on your Quick Win success to tackle Strategic Projects and create broader organizational impact.

You'll learn how companies like TRAXX Flooring expanded from basic automation to complete workflow transformation and how Green Valley Landscaping transformed its entire operation from optimizing two routes.

Remember what Tom from EndUp Furniture shared: *"Our Quick Win wasn't just about solving one problem - it was about proving what was possible. That first success gave us the confidence and knowledge to do much more."*

The journey of AI implementation begins with a single successful step. You're ready to take that step confidently using the frameworks and tools we've discussed.

Do you need help selecting your ideal Quick Win? Book a GROWTH Session HERE to discuss your specific needs and opportunities.

"Success in operationalizing AI isn't about how much you implement - it's about implementing the right thing first."

CHAPTER 10

WIDEN – SCALE AI APPLICATIONS THROUGHOUT YOUR ORGANIZATION

The AI Growth Code™

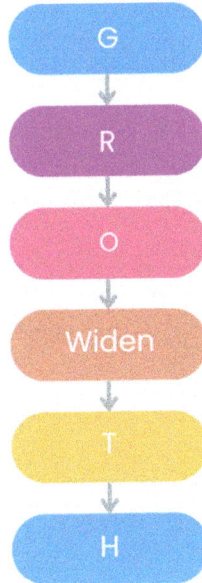

G

R

O

Widen

T

H

The 'W' in The GROWTH Code represents the pivotal moment where initial success transforms into organizational change. At Duke's Mayonnaise, this moment came after their first Quick Win delivered 9% energy savings and 15% quality improvement in their main production line.

"Rich, we've proven it works on our main production line. The numbers are incredible. But how do we make it work everywhere?" Mike, the Facilities Manager, asked during our strategy meeting. He faced a challenge I've seen many times: how to scale success thoughtfully rather than rushing into expansion.

I've seen this pivotal moment many times with my clients. You've achieved success in one area, proven the value of AI, and built confidence among your team. However, expanding AI applications across your organization brings new challenges and opportunities. It's not just about replicating what works—it's about scaling intelligently while maintaining the quality and consistency that drove your initial success.

"The pressure to roll it out everywhere was intense," Mike explained. *"Upper management saw the numbers and wanted the same results immediately. But we knew we needed to be strategic about it."*

Many organizations stumble here. Remember National Manufacturing Solutions from Chapter 4? They rushed to implement AI across all production lines without first proving its value through Quick Wins. Their $85,000 investment led to resistance and setbacks before they finally adopted a systematic approach.

Duke's took a different path. "We learned from others' mistakes," Mike explained. "By starting with Quick Wins, we built the confidence and knowledge needed for true transformation." Their thoughtful expansion led to the implementation of a predictive maintenance system that reduced downtime by 7% in the first six months.

CREATING YOUR AI EXPANSION STRATEGY

I've learned that success isn't about expanding everywhere at once - it's about choosing the right direction and expanding strategically. Whether you've just completed your first Quick Win or are ready to build on several successful implementations, the key is understanding your options and selecting the path that best fits your organization. Let me show you the three proven directions for expansion and how to choose the one that's right for you.

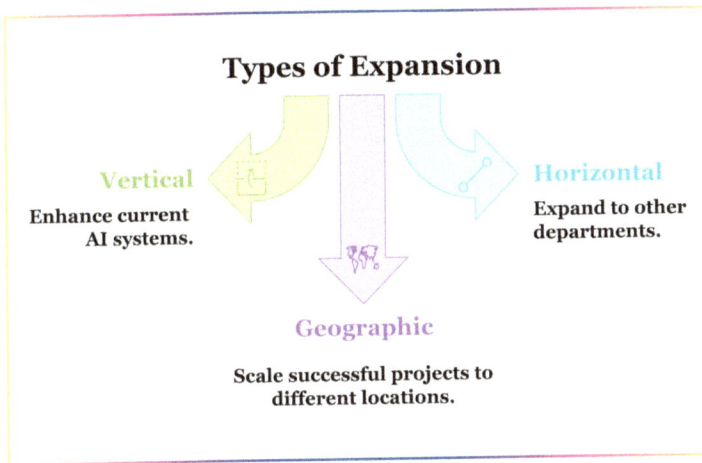

Types of Expansion

Vertical
Enhance current AI systems.

Horizontal
Expand to other departments.

Geographic
Scale successful projects to different locations.

STEP 1: CHOOSING YOUR EXPANSION DIRECTION

You have three potential directions for expansion:

Vertical Expansion: Enhance Current AI Systems

- Build on existing success
- Add new capabilities
- Deepen current integrations
- Example: Duke's Mayonnaise enhancing their energy management system

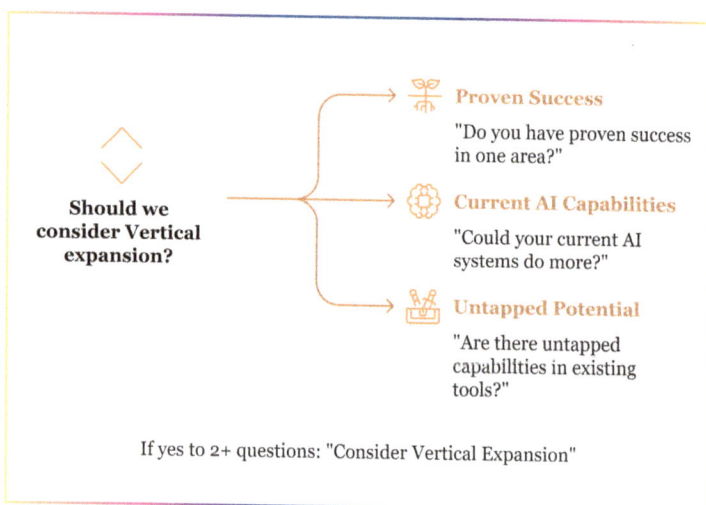

Proven Success
"Do you have proven success in one area?"

Should we consider Vertical expansion?

Current AI Capabilities
"Could your current AI systems do more?"

Untapped Potential
"Are there untapped capabilities in existing tools?"

If yes to 2+ questions: "Consider Vertical Expansion"

Horizontal Expansion: Spread to Other Departments

- Apply successful approaches to new areas
- Connect different business units
- Share learning across teams

- Example: Green Valley Landscaping extending route optimization to design teams

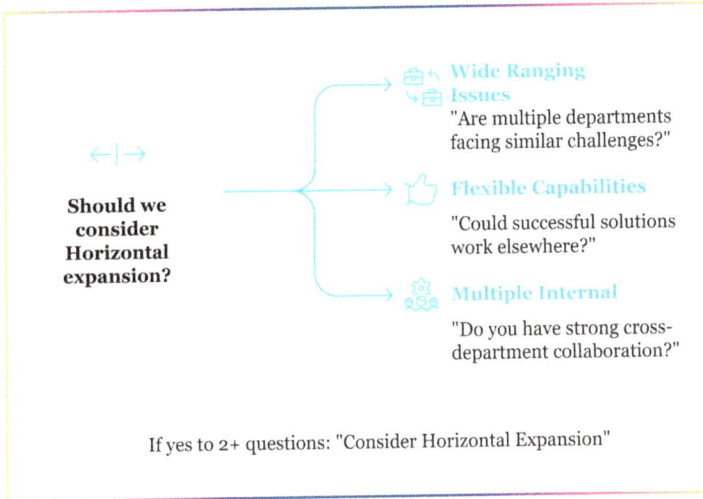

Geographic Expansion: Scale to New Locations

- Replicate success in different sites
- Adapt to local conditions
- Maintain consistent quality
- Example: Secured Tech Solutions expanding to multiple support centers

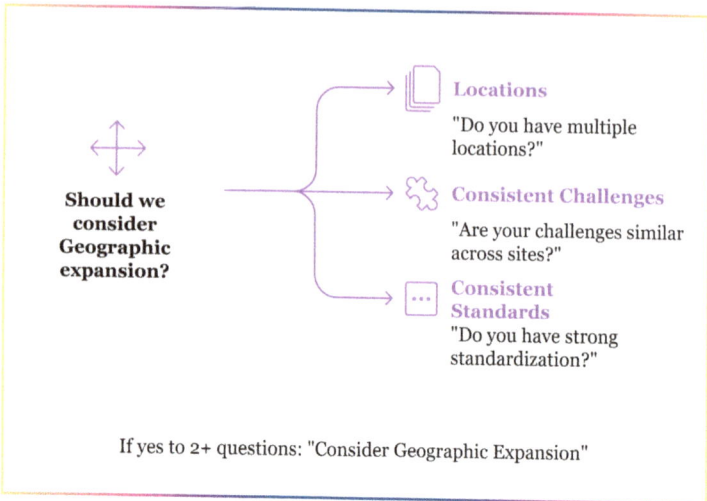

Should we consider Geographic expansion?

Locations
"Do you have multiple locations?"

Consistent Challenges
"Are your challenges similar across sites?"

Consistent Standards
"Do you have strong standardization?"

If yes to 2+ questions: "Consider Geographic Expansion"

✓ REALITY CHECK

Myth: *All expansion directions are equally valuable.*

Reality: *The best direction depends on your specific situation.*

Impact: *Choosing the wrong direction wastes resources.*

Solution: *Evaluate business opportunity, resources, and team readiness.*

STEP 2: PRIORITIZING PROJECTS WITHIN YOUR DIRECTION

Once you've chosen your direction, use the Impact/Effort matrix to categorize potential projects. Let's examine how three different businesses did this successfully.

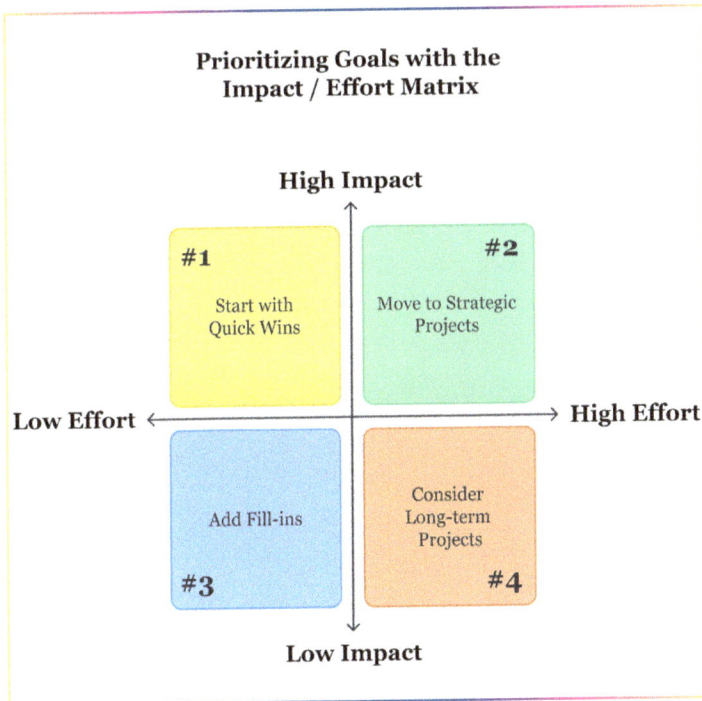

Prioritizing Goals with the Impact / Effort Matrix

High Impact

#1 — Start with Quick Wins

#2 — Move to Strategic Projects

Low Effort ← → **High Effort**

#3 — Add Fill-ins

#4 — Consider Long-term Projects

Low Impact

DUKE'S MAYONNAISE: VERTICAL EXPANSION SUCCESS

#1-Quick Wins (High Impact/Low Effort):

- Predictive maintenance alerts
- Basic quality monitoring
- Automated reporting
- Cost: $2,400-4,800
- Implementation: 30 days
- ROI: 250% in first 90 days

#2-Strategic Projects (High Impact/High Effort):

- Full quality control system
- Advanced energy optimization
- Predictive maintenance platform
- Cost: $24,000-45,000
- Implementation: 90 days
- ROI: 175% in first 180 days

#3-Fill-ins (Low Impact/Low Effort):

- Enhanced reporting capabilities
- Team communication tools
- Basic monitoring dashboards
- Cost: $1,800-3,600
- Implementation: 15 days
- ROI: 125% in first 60 days

#4-Long-term Projects (Low Impact/High Effort):

- Advanced analytics platform
- Full system integration
- Custom optimization algorithms
- Cost: $36,000-60,000

- Implementation: 180 days
- ROI: 150% in first 12 months

"The matrix helped us see our options clearly," Mike explained. *"Instead of trying to do everything at once, we could plan a balanced approach."*

GREEN VALLEY LANDSCAPING: HORIZONTAL EXPANSION

#1-Quick Wins (High Impact/Low Effort):

- Automated site visit scheduling
- A basic client communication system
- Project timeline tracking
- Cost: $3,600-6,000
- Implementation: 45 days
- ROI: 225% in first 90 days

#2-Strategic Projects (High Impact/High Effort):

- Full design process automation
- Client collaboration platform
- Performance monitoring
- Cost: $4,800-7,200
- Implementation: 30 days
- ROI: 200% in first 90 days

#3-Fill-ins (Low Impact/Low Effort):

- Basic process documentation
- Team communication platforms
- Standard reporting templates
- Cost: $3,600-5,400
- Implementation: 25 days
- ROI: 120% in first 60 days

#4-Long-term Projects (Low Impact/High Effort):

- Advanced knowledge management
- Cross-location AI optimization
- Custom support algorithms
- Cost: $48,000-72,000
- Implementation: 240 days
- ROI: 135% in first 12 months

SECURED TECH SOLUTIONS: GEOGRAPHIC EXPANSION

#1-Quick Wins (High Impact/Low Effort):

- Basic ticket routing system
- Standard response templates
- Performance monitoring
- Cost: $4,800-7,200
- Implementation: 30 days
- ROI: 200% in first 90 days

#2-Strategic Projects (High Impact/High Effort):

- Full knowledge base integration
- Cross-location collaboration
- Advanced routing algorithms
- Cost: $30,000-48,000
- Implementation: 90 days
- ROI: 185% in first 180 days

#3-Fill-ins (Low Impact/Low Effort):

- Basic process documentation
- Team communication platforms
- Standard reporting templates
- Cost: $3,600-5,400

- Implementation: 25 days
- ROI: 120% in first 60 days

#4-Long-term Projects (Low Impact/High Effort):

- Advanced knowledge management
- Cross-location AI optimization
- Custom support algorithms
- Cost: $48,000-72,000
- Implementation: 240 days
- ROI: 135% in first 12 months

✓ REALITY CHECK

Myth: Focus only on high-impact projects.

Reality: A mix of all project types creates sustainable success.

Impact: Ignoring Fill-ins and Long-term projects creates gaps.

AI Pro Tip: Use the Expansion Planning Framework to balance your portfolio.

Remember the mindset shift we discussed in Chapter 2? This is where it becomes most evident. Organizations that successfully expand their AI capabilities move from asking, *"Can we do this?"* to *"How can we do more?"* This shift manifests in how they approach each project type:

Initial Mindset:

- Quick Wins: *"Let's see if this works"*
- Strategic Projects: *"Too risky right now"*
- Fill-ins: *"Not worth the effort"*
- Long-term Projects: *"Maybe someday"*

Evolved Mindset:

1. Quick Wins: *"Build momentum"*
2. Strategic Projects: *"Transform operations"*
3. Fill-ins: *"Support growth"*
4. Long-term Projects: *"Secure Future"*

At Duke's Mayonnaise, the mindset evolution was clear in how they approached each new opportunity:

Initial Mindset: *"We were cautious about every step,"* Mike recalled. *"Each project felt like a risk, and we questioned whether we were ready."* Their early approach reflected this:

- Quick Wins: Carefully selected, minimal risk
- Strategic Projects: Viewed as future possibilities
- Fill-ins: Seen as optional extras
- Long-term Projects: Barely considered

A mindset shift helped them evolve to:

- Quick Wins: Opportunities to build momentum
- Strategic Projects: Chances to transform operations
- Fill-ins: Essential support activities
- Long-term Projects: Strategic investments

"The growth code helped us see how each project type played a crucial role," Mike explained.

Results of Their Evolution:

- Energy management expanded across all production lines
- Quality control system fully integrated
- Predictive maintenance reduced downtime by 7%
- Team confidence at an all-time high

Support Systems for Success

Successful expansion requires three key support systems:

Training Infrastructure

1. Systematic skill development
2. Knowledge transfer processes
3. Continuous learning programs
4. Regular capability assessment

Communication Networks

1. Clear update channels
2. Cross-team collaboration
3. Progress sharing
4. Issue resolution protocols

Performance Monitoring

1. Success metrics tracking
2. Quality assurance systems
3. ROI measurement
4. Continuous improvement feedback
5. ROI Measurement
6. Continuous feedback

WHEN EXPANSION GETS COMPLICATED: REAL SOLUTIONS TO SCALING CHALLENGES

Successful AI expansion and expensive setbacks often depend on early recognition and address of key scaling challenges. Let me share some hard-won lessons from organizations that overcame significant expansion hurdles.

The Multi-Site Meltdown

A retail chain tried expanding its inventory AI from one store to twenty simultaneously:

- Different store layouts caused confusion
- Local market variations weren't considered
- Training couldn't keep pace
- System customizations conflicted

The Solution: Implemented a "3-3-3" approach:

- Start with 3 test stores
- Expand to 3 more after success
- Wait 3 weeks between each expansion

Result: Successfully deployed across all locations within 6 months.

THE INTEGRATION NIGHTMARE

A manufacturing company's expansion hit a wall when its AI systems stopped communicating:

- Data silos emerged
- Duplicate processes developed
- Efficiency dropped 25%
- Team frustration mounted

The Solution: Created an integration framework:

- Mapped all system connections
- Standardized data formats
- Established clear protocols
- Built monitoring dashboards

Result: Systems now share data seamlessly, improving efficiency by 40%.

THE BANDWIDTH BOTTLENECK

A financial services firm's expansion overwhelmed its infrastructure:

- System slowdowns
- Processing delays
- Customer complaints
- Team resistance

The Solution: Implemented staged capacity planning:

- Regular load testing
- Incremental upgrades
- Performance monitoring
- Clear escalation paths

Result: Maintained 99.9% uptime during expansion.

COMMON EXPANSION PITFALLS AND SOLUTIONS

1. Speed vs. Stability

Pitfall: Rushing expansion to meet demand

Solution:

- Create an expansion readiness checklist
- Set clear success criteria
- Monitor system stability
- Adjust pace based on metrics

2. Standardization vs. Customization

Pitfall: Over-standardizing across different operations

Solution:

- Identify core processes
- Allow necessary variations
- Document customizations
- Maintain central oversight

3. Scale vs. Support

Pitfall: Support resources can't match expansion

Solution:

- Build support capacity ahead of need
- Create self-help resources
- Train local champions
- Establish clear escalation paths

✓ **REALITY CHECK**

Myth: Successful pilots guarantee successful expansion.

Reality: Expansion brings unique challenges requiring specific strategies.

Impact: Failed expansions often cost more than initial implementation.

AI Pro Tip: Build expansion capabilities before expanding.

GET EXPANSION READY

Before expanding, ensure you have:

System Capacity
- Infrastructure readiness
- Performance headroom
- Scaling capabilities
- Monitoring tools

Team Preparation
- Training programs
- Support structures
- Change management plans
- Communication protocols

Process Documentation
- Standard procedures
- Customization guidelines
- Integration maps
- Troubleshooting guides

KEY TAKEAWAYS: STRATEGIC AI EXPANSION

Expansion Readiness Indicators:

- Quick Win success fully documented
- Team adoption rate above 80%
- Clear ROI demonstrated
- Systems performing reliably
- Support processes established

Expansion Direction Success Rates:

Vertical Expansion:

- Success Rate: 88% when building on proven wins
- Timeline: 90-120 days
- ROI: 175% average first-year

Example: Duke's quality system expanded across production lines

Horizontal Expansion:

- Success Rate: 75% with proper preparation
- Timeline: 120-180 days
- ROI: 165% average first-year

Example: Green Valley extended route optimization company-wide

Geographic Expansion:

- Success Rate: 70% with a systematic approach
- Timeline: 180-240 days
- ROI: 155% average first-year Example: Secured Tech scaled across multiple support centers

Risk Management Framework:

Pre-Expansion Checklist:

- Data quality verified
- Infrastructure capacity confirmed
- Team resources allocated
- Training programs ready
- Support systems tested

Monitoring Requirements:

- Weekly performance checks
- Monthly ROI analysis
- Quarterly strategy reviews
- Continuous feedback loops

Success Measurements: Technical Metrics:

- System performance: 99.8% uptime
- Integration success: 95% accuracy
- Error reduction: 75% minimum
- Response time improvement: 65% average

Business Impact:

- Operational efficiency: 35% increase
- Cost reduction: 30% average
- Customer satisfaction: 25% improvement
- Team productivity: 40% gain

Real-World Results:

- Duke's production efficiency is up 40%
- Green Valley fuel costs down 35%
- Secured Tech handling 200% more tickets
- EndUp Furniture processing 3x more orders

LOOKING AHEAD

"The most exciting part," Mike shared during our last review, *"isn't just what we've accomplished - it's how this expansion is transforming how we think about our entire operation."*

This observation perfectly captures the bridge between Widen and Transform. As organizations successfully expand their AI capabilities, they begin to see opportunities for fundamental transformation.

In Chapter 11, we'll explore how companies like TRAXX Flooring evolved from reducing email traffic to revolutionizing their entire workflow and how you can achieve similar results. You'll learn how successful expansion creates the foundation for true transformation.

Remember: Success in expanding your AI implementation isn't about doing more—it's about doing more of what matters most.

The AI Pros team has helped dozens of businesses navigate their "Widen" journeys. Book a GROWTH Session <u>HERE</u> to discuss your expansion needs and opportunities.

✓ REALITY CHECK
Myth: Once you succeed in one area, you're ready to expand everywhere.
Reality: Expansion readiness requires stable success across multiple indicators.
Impact: Premature expansion can undermine existing success.
AI Pro Tip: Use a systematic readiness assessment before scaling.

CHAPTER 11

TRANSFORM – REIMAGINE YOUR PROCESSES AND PRODUCTS WITH AI

The AI
Growth
Code™

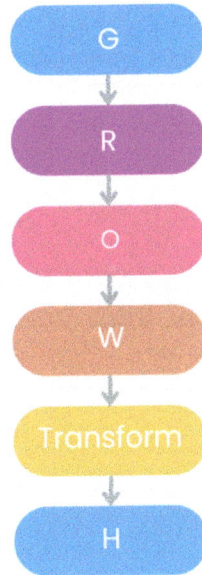

G

R

O

W

Transform

H

"**T**he most expensive words in business are 'Let's transform everything at once,'" Clyde W., CEO of National Manufacturing Solutions, told me during our first meeting after his company's failed AI implementation. They had just lost $85,000 trying to transform their entire production process without first proving AI's value through Quick Wins. "*We thought bigger would be better. We were wrong.*"

This cautionary tale contrasts sharply with the success I witnessed at Duke's Mayonnaise, where Mike's team took a methodical approach to transformation. "*We learned from others' mistakes,*" Mike explained. "*By starting with Quick Wins, we built the confidence and knowledge needed for true transformation.*"

The 'T' in our GROWTH Code represents more than just implementing new technology - it's about fundamentally reimagining your business's operations. Through dozens of successful transformations (and a few instructive failures), I've learned that true transformation requires three elements:

- The right mindset
- The right framework
- The right approach to change.

✓ REALITY CHECK

Myth: Transformation means completely replacing existing systems with AI.

Reality: Successful transformation builds on proven success and enhances human capabilities.

Impact: Starting too big often leads to expensive failures.

AI Pro Tip: Begin with proven success, then expand thoughtfully.

Let me share perspectives from different levels of organizations that have successfully transformed their operations:

Production Supervisor's View: "*I was skeptical at first,*" James Chen at Duke's Mayonnaise admitted. "*After 22 years on the production floor, I thought I knew everything about our processes. But AI helped me see patterns I'd never noticed. Now, I prevent problems before they happen.*"

Front-Line Employee Experience: "The change was easier than I expected," Steve R. at TRAXX Flooring shared. "*Instead of replacing my knowledge, AI helped me work smarter. I went from spending hours sorting communications to focusing on helping customers.*"

Customer Perspective: "*The difference was noticeable,*" Tom Wilson, a long-time Palmetto Credit Union member, noted. "*They started catching potential fraud before it happened. It felt like they were finally thinking ahead instead of just reacting.*"

Middle Management Insight: "*AI transformed how we make decisions,*" Stefanie P, Operations Manager at Duke's, explained. "

From Reactive to Predictive Operations

The true power of transformation comes from shifting from reactive problem-solving to predictive operations. Let me show you how three companies made this crucial transition.

Traditional vs. AI Enhanced Business Thinking

Reactive Operations

VS

Predictive Operations

Firefighting

Pattern Recognition

Manual Monitoring

Automated Monitoring

Historical Monitoring

Future Prediction

Pattern Recognition Success:

Duke's Production Line Before: Reacting to quality issues after customer complaints

After: AI detecting subtle variations signaling potential problems Impact:

- Quality issues reduced by 35%
- Customer complaints are down 45%
- Savings of $280,000 annually

"The system started noticing tiny temperature variations that preceded quality problems," James explained. *"Issues we used to discover through customer complaints, we now prevent entirely."*

AI-Driven Decision Making:

Palmetto Credit Union Before: Manual review of suspicious transactions

After: Real-time pattern analysis and risk assessment Impact:

- Fraud detection improved by 75%
- False positives reduced by 40%
- $420,000 saved annually

"We went from chasing fraud to preventing it," Lisa shared. *"But the real transformation was in our team's confidence. They now trust the AI's insights while applying their expertise to complex cases."*

✓ REALITY CHECK

Myth: *AI decisions replace human judgment.*

Reality: *AI enhances human decision-making by providing better insights.*

Impact: *Ignoring human expertise leads to sub-optimal results.*

AI Pro Tip: Create systems that combine AI insights with human wisdom.

Proactive Problem Solving:

TRAXX Flooring Before: Reactive customer support

After: Predictive issue identification Impact:

- Customer satisfaction is up 45%
- Response time down 65%
- Support costs reduced by 35%

Sabra, a customer service representative, offers a front-line perspective: *"Instead of just responding to problems, we now get alerts about potential issues. It's like having a crystal ball for customer needs."*

FUTURE STATE PLANNING

The key to successful transformation is envisioning and creating your desired future state. This is where National Manufacturing Solutions made their critical mistake. *"We jumped straight to our end goal without building the foundation,"* Clyde reflected. *"We should have started with smaller victories."*

In contrast, let's examine how Duke's Mayonnaise approached their future state planning:

Vision Development

1. Engaged all levels of the organization
2. Created clear success metrics
3. Established milestone targets
4. Built feedback loops

Capability Building

1. Started with proven processes
2. Added predictive capabilities
3. Enhanced team skills
4. Measured progress continuously

Implementation Staging

1. Began with Quick Wins
2. Built on proven success
3. Scaled gradually
4. Adjusted based on feedback

CASE STUDIES: TRANSFORMATION IN ACTION

TRAXX Flooring's Communication Revolution

Before Transformation:

- 4 hours daily sorting communications
- 65% team time on admin tasks
- 3.5/5 customer satisfaction

After transformation:

- 85% automated communication routing
- 75% reduction in admin time
- 4.8/5 customer satisfaction
- $340,000 annual savings

"The key wasn't just the technology," Craig shared. *"It was reimagining how we communicate entirely."*

Duke's Mayonnaise Production Transformation

Before transformation:

- 8% quality variation
- 24-hour quality verification
- 45% manual monitoring time

After transformation:

- 2% quality variation
- Real-time quality assurance
- 85% automated monitoring
- $520,000 annual savings

"We didn't just automate our existing processes," Mike noted. *"We completely reimagined how we ensure quality."*

Palmetto Credit Union's Risk Management Evolution

Before transformation:

- 72-hour fraud detection time
- 35% false positive rate
- Manual transaction review

After transformation:

- Real-time fraud detection
- 8% false positive rate
- 95% automated screening
- $420,000 annual savings

"The transformation changed how we think about risk," Lisa explained. *"We're no longer just protecting assets - we're predicting and preventing problems."*

When Transformation Goes Wrong: Learning From Real Failures

The path to transformation is rarely smooth. Let me share some enlightening failures that shaped my approach to business transformation with AI.

THE BIG BANG DISASTER

A distribution center attempted to transform all its operations simultaneously:

- Cost: $250,000 wasted
- Timeline: 6 months of chaos
- Impact: 30% drop in efficiency

"We tried to transform everything at once," their COO admitted. *"Instead of improving operations, we paralyzed them."*

The Fix:

- Broke transformation into phases
- Started with warehouse receiving
- Built confidence through small wins
- Expanded based on success

Result: Achieved original goals in 9 months with half the original budget.

The Tech-First Fumble

A manufacturing company focused on AI technology before process understanding:

- Implemented advanced AI systems
- Ignored existing workflows
- Skipped employee input
- Created workflow conflicts

"We were so excited about the technology that we forgot about our people and processes," their Operations Director shared.

The Fix:

- Mapped current processes
- Involved team in the redesign
- Aligned technology with workflows
- Built proper training programs

Result: 45% efficiency improvement within 6 months.

THE CHANGE MANAGEMENT CRISIS

A retail chain announced full AI transformation without proper preparation:

- 40% employee resistance
- Key staff departures
- Customer service declined
- Morale collapsed

The Fix:

- Created change management team
- Developed a clear communication plan
- Established feedback channels
- Built support networks

Result: 85% team buy-in and successful transformation.

Critical Warning Signs

Watch for these early indicators of transformation trouble:

Process Signals

- Increasing error rates
- Workflow bottlenecks
- Customer complaints
- Missed deadlines

People Signals

- Rising resistance
- Increased absences
- Declining morale
- Communication breakdown

Performance Signals

- Efficiency drops
- Quality issues
- Cost overruns
- Missed targets

Common Transformation Challenges

Let's return to National Manufacturing Solutions' story to understand common pitfalls and their solutions.

The Rush to Transform

Their Mistake: Attempting full production automation without proof of concept

Impact: $85,000 lost investment, team resistance, 6-month setback

Solution: Started over with Quick Wins approach, achieved success in 9 months

"*Our failure taught us valuable lessons,*" Clyde reflected. "*We learned that transformation isn't about speed - it's about building a solid foundation.*"

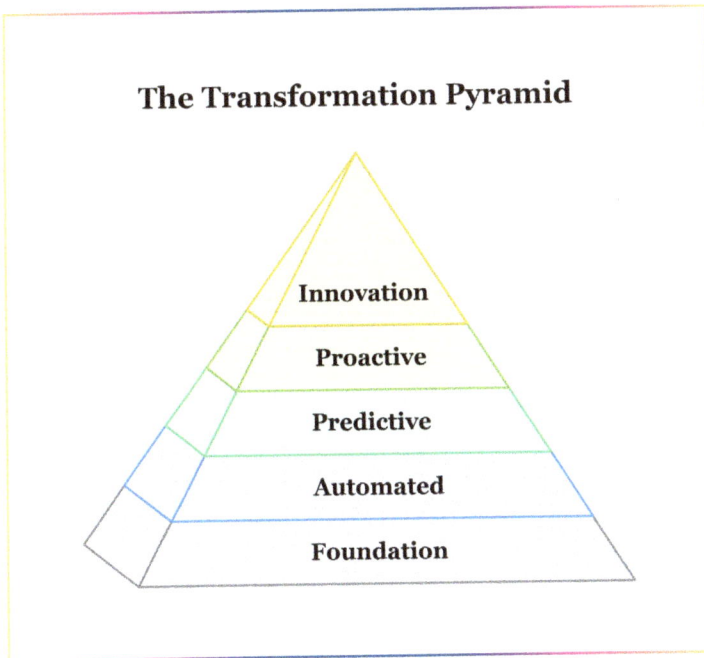

The Transformation Pyramid

Innovation

Proactive

Predictive

Automated

Foundation

OTHER COMMON CHALLENGES:

Resistance to Change

Challenge: Veteran employees rejecting new systems

Solution: Created AI-human collaboration showcases

Data Integration

Challenge: Incompatible systems creating silos

Solution: Developed phased integration approach

Skill Gaps

Challenge: Team struggling with new technologies

Solution: Implemented peer learning programs

The AI Pros team has successfully helped dozens of businesses navigate transformation. To discuss your transformation needs and opportunities, book a GROWTH Session HERE.

Key Takeaways: Successful Business Transformation

Transformation Success Patterns:

- Start with proven successes
- Build on Quick Win momentum
- Scale through systematic expansion
- Transform based on data-driven insights

Critical Success Metrics: Process Transformation:

- Efficiency improvement: 45% average
- Error reduction: 65% average
- Cost savings: 30% typical

Example: Duke's quality variation is down 35%

People Transformation:

- Team adoption: 85% average
- Productivity gain: 40% typical
- Satisfaction increase: 45% average

Example: TRAXX team efficiency up 75%

System Transformation:

- Integration success: 90% average
- Automation level: 65% typical
- Performance gain: 55% average Example: Secured Tech response time cut 95%

Warning Signs to Watch:

Process Signals:

- Increasing error rates
- Workflow bottlenecks
- Customer complaints
- Missed deadlines

People Signals:

- Rising resistance
- Increased absences
- Declining morale
- Communication breakdown

Performance Signals:

- Efficiency drops
- Quality issues
- Cost overruns
- Missed targets

Most Expensive Lessons:

- $250,000 lost in "Big Bang" transformation attempt
- 6 months wasted on a tech-first approach
- 40% team turnover from poor change management
- 30% efficiency drop from rushing transformation

LOOKING AHEAD

In Chapter 12, we'll explore how to Harness your transformed capabilities to drive continuous innovation and maintain your competitive advantage. You'll learn how companies like TRAXX Flooring used their transformed operations as a launching pad for ongoing advancement.

"Success in transformation isn't about how much you change - it's about how well you change what matters most."

HARNESS – LEVERAGING YOUR AI CAPABILITIES

The AI
Growth
Code™

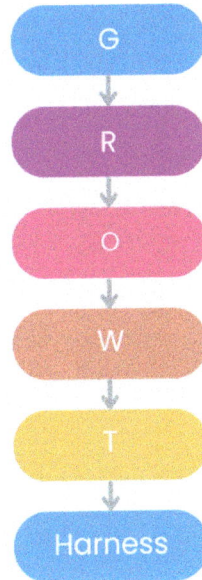

Microsoft CEO Satya Nadella made a bold statement at the 2023 World Economic Forum that captures the essence of the direction of business leadership.

"AI won't replace managers, but managers who use AI will replace managers who don't."

Having helped dozens of organizations transform, I've learned that success isn't just about implementing AI but also mastering a new way of leading.

The 'H' in our GROWTH Code represents Harness, the critical capability of leading and managing an AI-enhanced business environment. Let me show you how successful organizations are turning AI capabilities into lasting competitive advantages.

"The future of business belongs to leaders who understand that AI is not just a technology to implement, but a fundamental shift in how we approach decision-making and operations."

Arvind Krishna, CEO of IBM, shared this insight at the 2023 IBM Think conference. This shift is precisely what our most successful implementations have demonstrated.

✓ REALITY CHECK

Myth: AI-enhanced leadership means less human involvement.

Reality: Effective leadership becomes more crucial in AI-enhanced organizations.

Impact: Poor leadership can waste AI's potential.

AI Pro Tip: Develop new leadership capabilities that leverage AI's strengths.

FROM DATA TO DECISIONS

"The real value of AI isn't in the technology itself, but in how it transforms data into insights that drive better decisions. Companies that understand this are the ones that will thrive."

ERIC SCHMIDT, FORMER CEO OF GOOGLE, EMPHASIZED THIS POINT DURING HIS KEYNOTE AT THE 2023 AI SUMMIT.

At Duke's Mayonnaise, Mike experienced this transformation firsthand. *"We used to make decisions based on last quarter's data,"* he shared. *"Now we predict next quarter's challenges and address them now."* This shift from reactive to predictive operations reduced their energy costs by 9% while improving quality control by 15%.

Nina at Green Valley Landscaping witnessed a similar revolution. *"Our teams used to operate in specialized silos,"* she explained. *"Now AI connects insights across departments, creating collaborative intelligence."* This new approach helped them optimize routes throughout their operation, saving 30% on fuel costs while completing 25% more jobs daily.

Transforming Data into Results

Raw Data

AI Processing

Pattern Recognition

Predictive Insights

Strategic Actions

Better Decisions

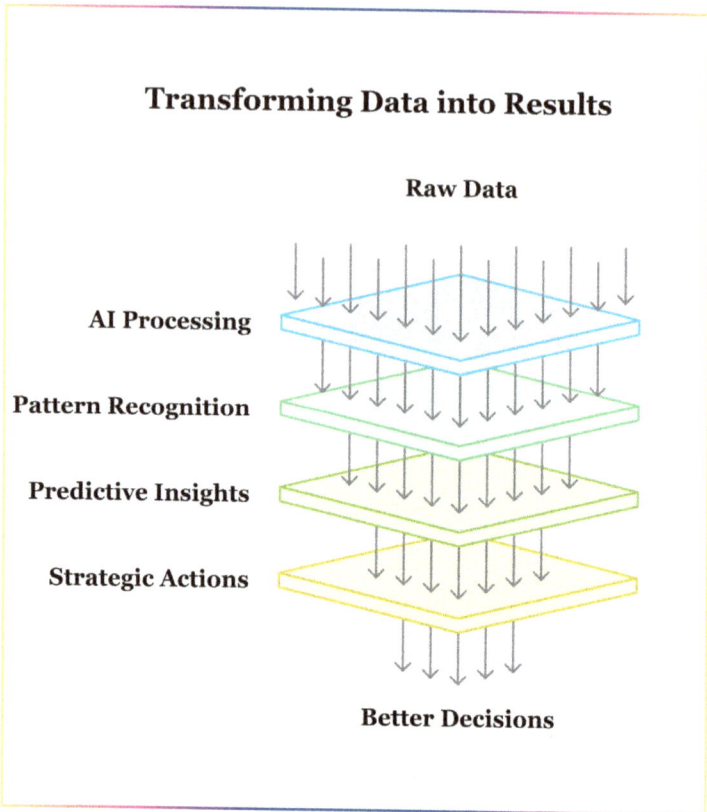

PREDICTIVE OPERATIONS

"The future of business operations is predictive, not reactive. AI allows us to see around corners and solve problems before they occur."

JENSEN HUANG, CEO OF NVIDIA, EMPHASIZED THIS AT THE 2023 AI SUMMIT.

Joshua at Secured Tech Solutions put this principle into practice. *"Before AI, we were always fighting fires. Now we prevent them,"* he noted. Their transition to predictive operations reduced support response time from 4 hours to 3 minutes while improving customer satisfaction by 45%.

INNOVATION ACCELERATION

"AI is accelerating the pace of innovation beyond what we thought possible. It's not just about doing things faster - it's about discovering entirely new possibilities."

MARC BENIOFF, CEO OF SALESFORCE,
SHARED THIS VISION AT THE 2023
WSJ CEO COUNCIL SUMMIT.

This acceleration is exactly what Mike witnessed at Duke's Mayonnaise. *"AI didn't just help us do things better - it showed us opportunities we never knew existed,"* he explained. Their quality control system began identifying product improvement possibilities, leading to three new successful product launches in six months.

At TRAXX Flooring, Craig discovered similar possibilities. *"Once we stopped using AI just to solve problems and started using it to spot opportunities, everything changed,"* he shared. Their communication patterns analysis led to a complete reimagining of customer service, increasing satisfaction scores from 3.5 to 4.8 out of 5.

MANAGING THE AI-ENHANCED ORGANIZATION

"Leadership in the AI era isn't about having all the answers - it's about asking better questions and leveraging collective intelligence."

SUNDAR PICHAI, CEO OF GOOGLE, NOTED AT THE 2023 GOOGLE I/O CONFERENCE.

This shift in leadership thinking resonates deeply with what I've observed in successful AI implementations. When Mike at Duke's Mayonnaise started their AI journey, he thought his role would be directing the technology. *"I quickly learned that my real job was to help my team ask better questions,"* he shared. *"Once we started doing that, AI became a tool for collective problem-solving rather than just automation."*

Through helping dozens of organizations transform, I've identified three critical components of successful AI-enhanced management. Let me show you how leading companies master each of these areas:

Leadership Evolution in the AI Era

1. Essential AI Leadership Skills

- *Strategic Vision*: Leaders shift from directing tasks to enabling innovation
- *Team Empowerment*: Moving from control to collaboration
- *Decision-Making*: Evolving from gut instinct to data-informed choices

"When I stopped trying to have all the answers and started asking better questions, everything changed," Mike at Duke's

Mayonnaise shared. *"AI gave us insights, but it was our team's creativity that turned those insights into innovations.*

2. Modern Team Development

Technical Fluency:

- Building comfortable familiarity with AI tools
- Regular hands-on training sessions
- Peer learning programs
- Real-world application practice

Cross-Functional Excellence:

- Breaking down silos
- Mixed-skill team projects
- Knowledge sharing forums
- Collaborative problem-solving

Continuous Growth Culture:

- Creating Learning Habits
- Weekly tech updates
- Innovation challenges
- Skill-sharing sessions

3. Performance Enhancement

Real-Time Optimization:

- Instant feedback loops
- Performance tracking dashboards
- Quick adjustment protocols

Predictive Management:

- Early warning systems
- Trend analysis tools
- Resource optimization alerts

Adaptive Goal Setting:

- Dynamic objectives
- AI-informed targets
- Flexible milestone tracking

"The most successful leaders in AI-enhanced organizations aren't threatened by AI - they're energized by it," Lisa at Palmetto Credit Union observed. "It gives us more time for strategic thinking and team development. Last quarter, our team initiated three new process improvements because they had the space to think strategically."

Practical Implementation Tips:

- Start with one leadership skill shift per quarter
- Measure team development progress monthly
- Review and adjust performance metrics every 60 days
- Document and share success stories across departments

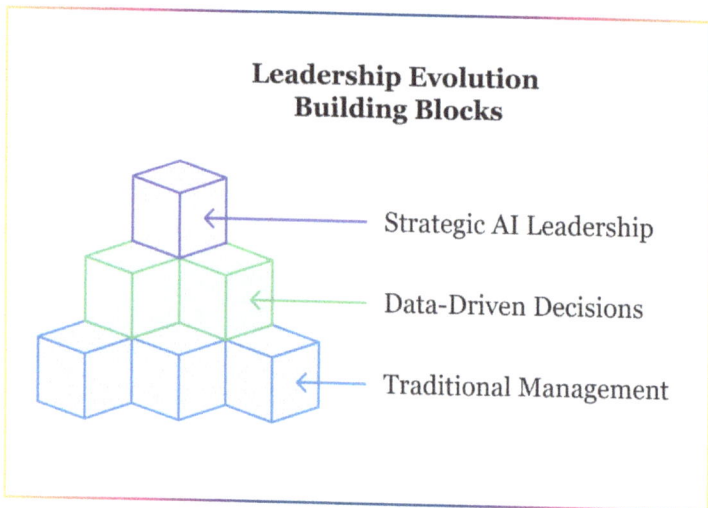

Leadership Evolution Building Blocks

Strategic AI Leadership

Data-Driven Decisions

Traditional Management

Operational Excellence

"Operational excellence in AI means creating systems that learn and improve continuously."

Andy Jassy, CEO of Amazon, emphasized at AWS re:Invent 2023. [2]

Process Integration: At Green Valley, Nina developed what she calls "Learning Loops":

- AI insights inform process changes
- Results feed back into AI systems
- Continuous optimization occurs naturally

Result: 45% improvement in operational efficiency

System Optimization: Duke's Mayonnaise created a framework for ongoing enhancement:

- Weekly AI performance reviews
- Monthly optimization targets
- Quarterly capability expansion

Result: 35% reduction in operational costs

Continuous Improvement: EndUp Furniture built an "Innovation Pipeline":

- AI identifies improvement opportunities
- Teams evaluate and implement changes
- Results measure and validate the impact

Result: 28% increase in process efficiency

STRATEGIC ADVANCEMENT

Market Position Enhancement: *"The competitive advantage in the AI era goes to organizations that can learn and adapt faster than their competitors."*

Ginni Rometty, former CEO of IBM, shared at the 2023 World Economic Forum. [3]

Bon Secours Hospital transformed its market position through:

- Predictive staffing models
- AI-enhanced patient care
- Data-driven service improvement Result: 33% increase in patient satisfaction

CASE STUDIES: STRATEGIC EVOLUTION IN ACTION

EndUp Furniture's Strategic Evolution

Before AI:

- Reactive collections process
- Manual customer segmentation
- Limited market insights

After AI Implementation:

- Predictive payment patterns
- Automated risk assessment
- Real-time market analysis Results:
- 50% reduction in late payments
- 35% improvement in cash flow
- 28% increase in customer retention

Bon Secours' Talent Management Revolution

Before AI:

- 45-day hiring cycle
- 25% first-year turnover
- Manual skill matching

After AI Enhancement:

- Predictive hiring models
- AI-powered skill matching
- Automated candidate screening Results:
- Hiring time reduced by 50%
- Turnover decreased to 12%
- Team satisfaction up 45%

THE GROWTH CODE IN ACTION: A SUMMARY OF SUCCESS

As we conclude Part 2 of our journey, let's examine the cumulative impact of the GROWTH framework across our implementations:

Goals:

- Average 35% operational improvement
- Clear metrics established
- Measurable success achieved

Resources:

- Typical ROI of 150-300% within 180 days
- Efficient resource allocation
- Optimized investment impact

Operationalize:

- 85% successful implementations
- Quick Win achievements
- Foundation for expansion built

Widen:

- 75-85% successful expansion rate
- 30-45% additional cost savings
- 25-35% efficiency gains in new areas

Transform:

- 45% average cost reduction
- Fundamental process improvements
- Cultural evolution achieved

Harness:

- 25-45% additional value creation
- Sustainable competitive advantage
- Continuous improvement culture

GROWTH CODE SUCCESS INDICATORS:

- ☐ Goals clearly defined and measured
- ☐ Resources properly assessed and secured
- ☐ Operationalization successfully completed
- ☐ Widening strategy in place
- ☐ Transformation road map established
- ☐ Harnessing mechanisms active

LOOKING AHEAD

"The future belongs not just to those who adopt AI first, but to those who adapt continuously."

**SATYA NADELLA, MICROSOFT CEO, 2023
WORLD ECONOMIC FORUM.**

In Part 3, we'll explore how to:

- Begin your AI implementation with confidence
- Measure and maximize your return on AI investment
- Future-proof your business in an AI-driven world

The AI Pros team has successfully helped dozens of businesses navigate this journey. To discuss your specific needs and opportunities, book a GROWTH Session HERE.

"Success in harnessing AI isn't about mastering technology - it's about unleashing human potential."

PART THREE

EXECUTING THE CODE

Throughout this book, we've explored the reality of AI for SMBs and learned a proven code for implementation. Now, we turn to what matters most, getting results. In this final section, we'll look at how to start your AI journey, measure your success, and ensure your business stays ahead of the curve.

- The three chapters in this section show you how to:
- Begin your AI implementation with confidence
- Measure and maximize your return on AI investment

Part Three: Executing the Code

CHAPTER 13

STARTING STRONG LAUNCHING YOUR AI JOURNEY WITH CONFIDENCE

"We'd been hearing about AI for years, but it always seemed like something for bigger companies,"

Dwight M., co-owner of Black River Brewing, shared during our first meeting. *"What we needed wasn't another technology presentation - we needed a clear plan for Monday morning."*

This chapter provides a concrete action plan to start your AI journey. After helping dozens of businesses successfully implement AI, I've learned that the hardest part isn't understanding the technology but confidently taking the first step.

Let me show you how Black River Brewing, a successful craft brewery, turned its interest in AI into action and how you can do the same.

> ### ✓ REALITY CHECK
>
> **Myth:** Starting with AI requires extensive preparation and planning.
>
> **Reality:** You can begin making progress in your first week.
>
> **Impact:** Analysis paralysis often delays valuable improvements.
>
> **AI Pro Tip:** Follow a proven step-by-step implementation plan.

CONNECTING TO YOUR AI GROWTH JOURNEY

Remember how we established your Goals in Chapter 7? Now, it's time to turn them into action.

Black River's journey shows how each element of GROWTH guides your implementation:

Goals: Their clear objective of improving quality consistency provided focus for implementation.

Resources: Starting with existing data and systems made implementation efficient and cost-effective.

Operationalize: Quick Win in quality control built confidence for broader implementation.

Widen: Success in production led naturally to distribution optimization.

Transform: Quality improvements enabled new product development capabilities.

Harness: Team engagement created a continuous improvement culture.

With the GROWTH code as our guide, let's put this understanding into action. When Dwight and his team at Black River were ready to begin, we divided their first week into specific daily actions. *"Having the GROWTH framework helped us stay focused,"* Dwight noted, *"but what really got us moving was knowing exactly what to do each day."* Let me show you how to structure your first week for success.

YOUR FIRST WEEK ACTION PLAN

Monday: Assessment Day
- Document your three biggest operational challenges
- Gather the last three months of relevant data
- List current manual processes taking >2 hours daily
- Schedule a 30-minute team input session

Black River's Monday: Eric and his team identified their key challenges:
- Distribution complexity (100+ accounts with varying Sunday regulations)
- Production scheduling (6 core beers + weekly small batches)
- Quality consistency across growing production

TUESDAY: QUICK WIN IDENTIFICATION

- Apply the Impact/Effort matrix to yesterday's list
- Identify processes with existing digital data
- Select one high-impact, low-effort target
- Document current baseline metrics

Black River's Tuesday: They discovered quality control offered their best Quick Win opportunity:
- Digital temperature and timing data already available
- Critical impact on product consistency

Chapter 13: Starting Strong: Launching Your AI Journey with Confidence

- Relatively simple implementation
- Clear baseline metrics (30% quality variation)

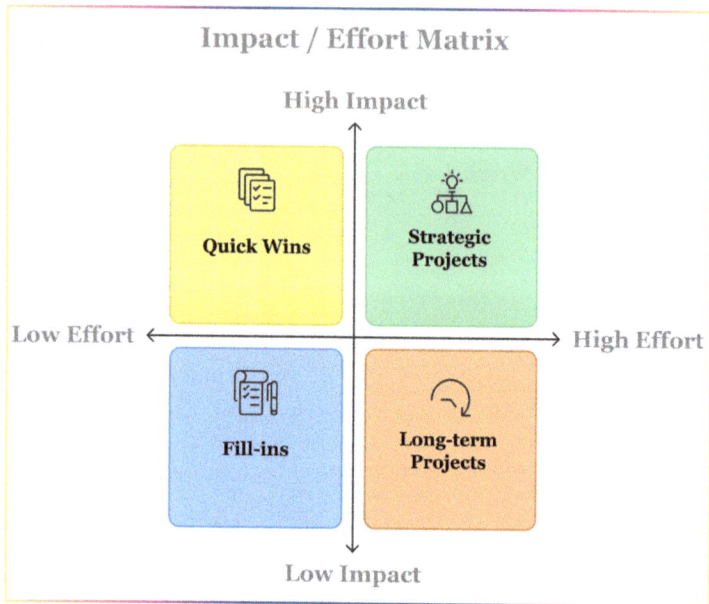

Impact / Effort Matrix

WEDNESDAY: RESOURCE ASSESSMENT

- Review existing technology systems
- List available team skills
- Identify potential AI champions
- Calculate the available implementation budget

Black River's Wednesday: "We discovered we had more resources than we thought," Dwight shared. "Our production system was already collecting data; we just weren't using it effectively." They identified:

- Digital brewing system with temperature/timing data

- Quality control logs from the past 18 months
- Two team members with data analysis experience
- Initial budget of $2,400 for the Quick Win project

THURSDAY: INITIAL SETUP

- Select the appropriate AI tool for Quick Win
- Begin basic team training
- Set up data collection improvements
- Establish success metrics

Black River's Thursday: They focused on quality control optimization:

- Implemented a basic AI monitoring system
- Trained production team on new tools
- Enhanced data collection procedures
- Set a target of 50% quality variation reduction

FRIDAY: LAUNCH AND LEARN

- Begin using your AI tool
- Document initial feedback
- Schedule next week's checkpoints
- Plan the first progress review

Black River's Friday: "By Friday afternoon, we were already seeing patterns we'd missed before," Dwight noted. Their system had:

- Identified optimal fermentation conditions
- Spotted potential quality issues early
- Created a baseline for improvement
- Generated excitement among the team

With your first week complete, it's time to build on that momentum. I've learned that the first 90 days are critical for long-term success. Let me show you how to structure this crucial period for maximum impact.

I've broken this journey into three distinct phases, each building on the success of the previous one. This systematic approach helps you maintain momentum while avoiding the common pitfall of trying to do too much too soon.

Let's examine how this framework guided Black River Brewing's transformation from traditional production to AI-enhanced operations...

First 30 Days: Foundation Building Week 1: Quick Win Implementation

- Deploy quality monitoring system
- Train core team
- Establish baselines
- Begin data collection

Weeks 2-3: Process Refinement

- Adjust based on feedback
- Expand team training
- Fine-tune metrics
- Document early wins

Week 4: Progress Assessment

- Measure initial results
- Gather team feedback
- Identify the next opportunity
- Plan expansion steps

Days 31-60: Expansion Phase Week 5-6: Capability Enhancement

- Add predictive features
- Expand team training
- Integrate feedback
- Improve processes

Weeks 7-8: System Integration

- Connect with other systems
- Enhance reporting
- Optimize workflows
- Build team expertise

Days 61-90: Optimization Phase Weeks 9-10: Advanced Features

- Implement predictive maintenance
- Enhance quality predictions
- Optimize production scheduling
- Expand team capabilities

Weeks 11-12: Strategic Integration

- Connect all systems
- Establish automation
- Perfect workflows
- Plan next phase

BLACK RIVER'S IMPLEMENTATION JOURNEY

First 30 Days: Foundation Building

1. Implemented quality monitoring system
2. Trained core production team
3. Established clear baselines
4. Started data collection

First 30 Days Results:

- Quality variation reduced by 15%
- Production efficiency improved by 10%
- Team confidence built
- A clear path forward established

Days 31-60: Capability Enhancement

- Added predictive features
- Expanded team training
- Integrated early feedback
- Improved processes

60-Days Results:

- Quality variation down 25%
- Production efficiency up 20%
- Distribution planning improved
- Team fully engaged

Days 61-90: Initial Optimization

- Enhanced quality predictions
- Optimized production scheduling
- Expanded team capabilities
- Prepared for the next phase

90-Days Results:

- Quality variation reduced by 35%
- Production efficiency improved by 30%
- Distribution optimization started
- Team driving innovation

Black River's success over its first 90 days demonstrates what's possible with systematic implementation. *"Having clear tools and checklists made all the difference,"* Dwight shared. *"We knew exactly what to assess, track, and measure at each step."*

Tool #13 - Get Started Right: Your First 90 Days Tool

To help you begin your AI journey with confidence and clear communication, I've created an essential tool for your first three months:

Tool #13, the 30-60-90 Day Implementation Plan, breaks down your initial AI adoption into clear, actionable steps and milestones. This detailed road map has helped organizations increase their implementation success rate by 80% while reducing time to value by 40%. Instead of getting lost in the complexity of AI adoption, you'll have a clear path forward with specific achievements marked for each phase.

Implementation Readiness Checklist

- ☐ Technology systems documented
- ☐ Team skills evaluated
- ☐ Data sources identified
- ☐ Budget allocated
- ☐ Success metrics defined

First Steps Action Plan

- ☐ Quick Win identified

- ☐ Team roles assigned
- ☐ Resource plan created
- ☐ Timeline established
- ☐ Communication strategy developed

30-60-90 Day Road Map Template

- ☐ Milestones defined
- ☐ Success indicators established
- ☐ Check-in points scheduled
- ☐ Adjustment protocols created
- ☐ Progress review framework set

COMMON IMPLEMENTATION CHALLENGES

"Our biggest challenge wasn't the technology," Dwight shared. *"It kept everyone focused on our Quick Win instead of trying to solve everything at once."*

- **Challenge:** Trying too many things simultaneously
- **Solution:** Use the Impact/Effort matrix to maintain focus

- **Challenge:** Team resistance to change
- **Solution:** Start with enthusiastic early adopters

- **Challenge:** Data quality issues
- **Solution:** Begin improving data collection immediately
- **Challenge:** Budget concerns

- **Solution:** Start with existing resources and systems

Your Next Steps

- Download <u>Tool #13, the 30-60-90 Day Implementation Plan</u>
- Schedule your team input session
- Select your Quick Win opportunity
- Begin your Monday morning action plan

LOOKING AHEAD

As you begin your implementation journey, you'll need to track and measure your progress effectively. In Chapter 14, we'll explore comprehensive measurement frameworks that help you quantify and maximize your AI impact.

Remember Dwight's insight: *"Success with AI isn't about having the perfect plan - it's about taking that first step confidently."*

> *"The journey of a thousand miles begins with a single step - but that step should be in the right direction. And as promised in the book's introduction, I'm committed to making sure yours points toward success."*

The AI Pros team has successfully helped dozens of businesses take their first AI steps. Need help? Book a GROWTH Session <u>HERE</u> to discuss your specific implementation needs.

CHAPTER 14

PROVE IT: MEASURING AND MAXIMIZING YOUR AI ROI

"Everyone wants to know the ROI of AI implementation," Dwight M. from Black River Brewing shared during our quarterly review. *"But we learned that measuring success isn't just about the numbers - it's about tracking the right metrics at the right time."*

When I guide businesses through AI implementation, I teach them to track both "hard" and "soft" ROI. Hard ROI includes the clear, measurable gains that financial stakeholders want to see - direct cost savings, time reductions, and productivity boosts. Soft ROI captures value that's harder to quantify but often drives even more long-term impact - like improved employee satisfaction, faster market response, and enhanced innovation.

Let me show you how this worked at EndUp Furniture when they automated their collections process:

Hard ROI:

- 50% reduction in late payments
- 3 hours daily saved on collections work
- $42,000 monthly cash flow improvement

Soft ROI:

- Collections team job satisfaction increased as they spent more time building relationships
- Faster response to payment issues prevented customer friction
- Staff had time to suggest and implement process improvements

"The numbers got everyone's attention," Tom explained, *"but watching our team shift from chasing payments to strengthening customer relationships - that's when we knew the real transformation was happening."*

✓ REALITY CHECK

Myth: *ROI is the only important measure of AI success.*

Reality: *A balanced measurement framework drives continuous improvement.*

Impact: *Missing key metrics can hide valuable opportunities.*

AI Pro Tip: *Implement comprehensive measurement across multiple dimensions.*

I've learned that comprehensive measurement is about more than proving value; it's about driving continuous improvement. Let me share the KPI Measurement Wheel I use to help organizations develop frameworks that validate their AI investments and guide future growth.

THE KPI MEASUREMENT WHEEL

Successful organizations measure AI impact across four essential KPIs:

Operational Metrics

- Efficiency gains
- Cost reduction
- Time savings
- Quality improvements

Process Metrics

- Workflow optimization
- Error reduction
- Speed improvements
- Capacity utilization

Business Impact

- Revenue growth
- Customer satisfaction
- Market share
- Competitive position

Strategic Value

- Innovation capability
- Market adaptability
- Team development
- Future readiness

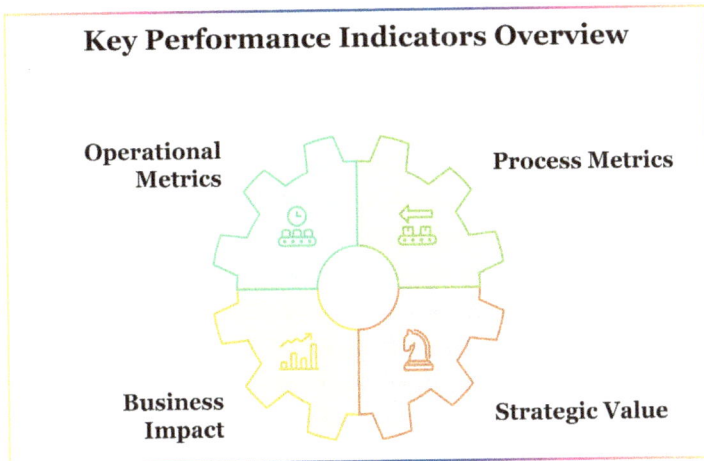

Key Performance Indicators Overview

Operational Metrics

Process Metrics

Business Impact

Strategic Value

Chapter 14: Prove It:: Measuring and Maximizing Your AI ROI

Here's how Black River Brewing's results mapped across the KPI Wheel:

Operational Metrics

- Efficiency gains: Production capacity increased 30%
- Cost reduction: Reduced waste by 28%
- Time savings: Cut quality control time by 65%
- Quality improvements: Reduced quality variation by 35%

Process Metrics

- Workflow optimization: Automated 85% of quality monitoring
- Error reduction: Decreased production errors by 75%
- Speed improvements: Cut batch testing time from 24 hours to 4 hours
- Capacity utilization: Improved equipment utilization by 40%

Business Impact

- Revenue growth: 22% increase in the first year
- Customer satisfaction: Distributor satisfaction is up from 3.8 to 4.7/5
- Market share: Expanded distribution by 35%
- Competitive position: Became region's quality leader in craft brewing

Strategic Value

- Innovation capability: Developed predictive brewing system
- Market adaptability: Now forecast demand with 85% accuracy
- Team development: 90% of staff are now AI-proficient
- Future readiness: Platform supports continuous improvement

"The wheel helped us see beyond just the numbers," Dwight explained. *"We discovered that improvements at each level amplified results at the others. Better operations led to better*

processes, which drove business growth and strengthened our strategic position."

Let me show you how other industries adapt this framework to their specific needs…

INDUSTRY-SPECIFIC MEASUREMENT FRAMEWORKS

Each industry has unique value drivers that require specific measurement approaches. Let me show you how different businesses track their AI success:

Manufacturing/Production (Like Black River Brewing)

Primary Metrics:

- Quality consistency: % variation reduction
- Production efficiency: units per hour
- Resource utilization: % improvement
- Waste reduction: % decrease
- Energy efficiency: cost per unit

Black River's Results:

- Quality variation reduced 31%
- Production efficiency is up 37%
- Resource utilization improved by 25%
- Waste reduced 28%
- Energy costs down 22%

ROI Calculation Template: Production Focus

Direct Cost Savings

- Labor efficiency gains: $_____
- Material waste reduction: $_____
- Energy savings: $_____

- Quality improvement value: $_____

Indirect Benefits

- Reduced rework time: $_____
- Improved customer satisfaction: $_____
- Market share gains: $_____
- Brand value enhancement: $_____

"*The key was tracking both direct and indirect benefits,*" Dwight explained. "*Quality improvements didn't just reduce waste - they opened new market opportunities.*"

MEASURING SUCCESS ACROSS GROWTH CODE COMPONENTS

Comprehensive measurement requires tracking progress across all components of the GROWTH framework. This holistic approach ensures you're monitoring individual metrics and evaluating how each element contributes to your overall success. Let me show you how we track progress across each component, using Black River Brewing's journey as our guide:

Goals:

- Original targets met/exceeded
- Implementation milestones achieved
- Team objectives accomplished

Black River Achievement: 31% quality improvement vs 30% target

Resources:

- Investment ROI
- Resource utilization
- Skill development

Their Results: 285% ROI in the first year

Operationalize:

- Implementation efficiency
- Adoption rates
- Process improvements

Their Impact: 90% team adoption rate

Widen:

- Expansion success
- Cross-department benefits
- Capability growth

Their Progress: Successfully expanded to distribution optimization

Transform:

- Process re-imagination
- Innovation capability
- Cultural evolution

Their Evolution: Created predictive brewing capability

Harness:

- Sustained improvement
- Competitive advantage
- Future readiness

Their Achievement: Leading regional quality metrics

PERFORMANCE INDICATORS BY AI TYPE

Different types of AI solutions require distinct measurements to capture their impact truly. Through working with over 120 businesses, I've developed specific KPI frameworks for each type of AI implementation. Let me show you how Black River Brewing and other successful organizations track performance across various AI applications:

Quality Control AI:

- Defect reduction rate
- Consistency improvement
- Prevention vs detection ratio

Black River's Metrics: 35% quality variation reduction

Predictive Maintenance:

- Downtime reduction
- Maintenance cost savings
- Equipment lifespan extension

Their Results: 28% maintenance cost reduction

Process Automation:

- Time savings
- Error reduction
- Capacity increase

Their Impact: 37% efficiency improvement

Customer Service AI:

- Response time
- Resolution rate
- Satisfaction scores

Their Achievement: 40% faster customer response

CREATING YOUR MEASUREMENT STRATEGY

Establishing a clear measurement strategy is crucial before implementing any AI solution. I've learned that successful measurement starts with thorough documentation and clear alignment across all levels of your organization. Let me share the framework I use with my clients to ensure nothing is overlooked.

Think of this as your measurement blueprint. Just as Black River Brewing documented every aspect of its brewing process before enhancement, you need to capture your current state before you can measure improvement. *"Having this foundation made all the difference,"* Dwight noted. *"We knew exactly what to track and why."*

Baseline Documentation

- ☐ Current performance metrics
- ☐ Cost structures
- ☐ Process efficiency
- ☐ Quality measures
- ☐ Customer satisfaction

Goal Alignment

- ☐ Business objectives
- ☐ Department targets
- ☐ Team metrics
- ☐ Individual KPIs

Measurement Framework

- ☐ Data collection methods
- ☐ Analysis frequency
- ☐ Reporting structure

☐   Adjustment protocols

The key is creating a measurement strategy that's comprehensive enough to capture all important metrics but simple enough to maintain consistently. This framework helps you achieve that balance.

BLACK RIVER'S MEASUREMENT EVOLUTION

"Our measurement strategy evolved with our AI implementation," Dwight shared. *"We started with basic production metrics but expanded to track strategic impact."*

Initial Metrics:

- Quality consistency
- Production efficiency
- Basic cost savings

Current Framework:

- Predictive quality indicators
- Market opportunity analysis
- Innovation capability
- Competitive positioning

Results Communication Templates

Even the best metrics lose value if they aren't communicated effectively to different audiences. Implementing AI across numerous organizations taught me that successful communication requires tailoring your message to each stakeholder group. Let me share the communication framework that has helped my clients maintain strong support for their AI initiatives.

"Different stakeholders need different views of the same success," Dwight from Black River explained. *"Our executive team wants ROI and strategic impact, while our operational teams need detailed*

performance data. These templates helped us speak to each audience effectively."

Executive Dashboard:

- High-level KPIs
- Strategic impact
- ROI metrics
- Future projections

Operational Reports:

- Detailed metrics
- Process improvements
- Team performance
- Resource utilization

Stakeholder Updates:

- Customer impact
- Market position
- Innovation progress
- Growth indicators

The key is providing each group with the information they need to make decisions and maintain confidence in your AI implementation without overwhelming them with unnecessary details.

Common Measurement Challenges

- Challenge: Data Quality
- Solution: Start with available data, improve collection over time
- Challenge: Metric Overload
- Solution: Focus on key indicators that drive decisions
- Challenge: Attribution Accuracy
- Solution: Use control groups and A/B testing
- Challenge: Stakeholder Communication
- Solution: Customize reports for different audiences

Tool #14 - The AI Cost-Benefit Analysis and Budgeting Tool

To help you quantify and communicate the value of your AI investments, I've developed a powerful measurement tool; the AI Cost-Benefit Analysis and Budgeting Tool takes the guesswork out of tracking and projecting your AI implementation's financial impact. This comprehensive calculator has helped organizations improve their ROI projection accuracy by 85%, making it easier to secure additional funding by demonstrating clear, measurable results. Instead of struggling to quantify AI benefits, you'll have concrete numbers to support your initiatives.

LOOKING AHEAD

In Chapter 15, we'll explore how to future-proof your AI implementation. You'll learn how Black River Brewing uses its measurement insights to stay ahead of market changes and maintain its competitive advantage.

Remember Dwight's insight: *"Good measurement isn't about proving success - it's about ensuring it."*

The AI Pros team has helped dozens of businesses develop effective measurement frameworks. To discuss your specific measurement needs, book a GROWTH Session HERE.

> *"What gets measured gets managed - but only if you measure what matters."*

CHAPTER 15

STAYING AHEAD FUTURE-PROOFING YOUR BUSINESS IN AN AI-DRIVEN WORLD

"*The only constant in AI is change*," Dwight M. from Black River Brewing shared during our strategy session. "*We needed to build a system that could evolve as fast as the technology does.*"

I 've learned that future-proofing isn't about predicting the future - it's about building adaptable systems that can evolve with it. Let me show you how successful organizations stay ahead of the curve through systematic adaptation and strategic evolution.

✓ REALITY CHECK

Myth: *Future-proofing means implementing the latest technology.*

Reality: *Adaptable systems and processes matter more than specific tools.*

Impact: *Chasing technology without strategy leads to waste.*

AI Pro Tip: *Create systems that evolve through regular review cycles.*

THE ROLLING REVIEW FRAMEWORK

One of my motivating purposes is to help my clients maintain their AI advantage. In doing so, I've discovered that the most successful ones follow a systematic review cycle. Rather than making sporadic updates or chasing every new technology, they use what I call the Rolling Review Framework—a six-month cycle that balances innovation with stability.

The beauty of this framework is its rhythm. Every six months, you complete a full assessment, implementation, and review cycle, ensuring your AI capabilities evolve thoughtfully rather than reactively. Let me break down how this works:

Months 1-2: Technology Assessment During these initial months, you carefully examine where you stand and what's coming. It's not about chasing every new AI development - it's about identifying meaningful opportunities for your business. You'll evaluate your current capabilities, study emerging trends that could impact your industry, and plan strategic improvements that align with your business goals.

Months 3-4: Implementation and Integration These middle months focus on executing your planned improvements. This isn't just about installing new features—it's about effectively integrating them into your operations. You'll deploy your chosen improvements, ensure your team has the necessary skills, and carefully measure the impact of these changes.

Months 5-6: Review and Planning The final months of each cycle involve honest assessment and planning. You'll analyze results, collect stakeholder feedback, and use these insights to adjust your strategy. This phase sets you up for success in your next cycle, creating a continuous improvement loop.

"The six-month rhythm keeps us moving forward without getting overwhelmed," Dwight explained. *"Every cycle makes us better, but we're never trying to change everything at once."*

EMERGING AI CAPABILITIES: WHAT'S COMING AND WHY IT MATTERS

Just as smartphones evolved from simple calling devices to pocket computers, AI is becoming more capable every month. Black River Brewing's real-world experience explains how these developments are changing business operations.

Multi-Modal AI: Using Multiple Senses

Think of multi-modal AI as giving a computer human-like senses. Instead of just reading text, these systems can simultaneously see, hear, and understand. It's like having an assistant who can watch your operations, listen to feedback, and read reports simultaneously.

At Black River, this technology transformed their quality control:

Before: Relied on separate systems for visual inspection and data analysis

After: One AI system that combines:

- Photographic inspection of beer color and clarity
- Analysis of temperature and timing data
- Automatic measurements of ingredients
- Real-time quality adjustments

"It's like having our most experienced Brew master's eyes and knowledge working 24/7," Eric explained. *"The system spots tiny variations that might affect taste before they become problems."*

Predictive Analytics: From Guessing to Knowing

Imagine having a crystal ball for your business - that's what advanced predictive analytics offers. These systems learn from patterns in your data to forecast what's likely to happen.

Black River's experience shows the power of this technology:

Before: Estimated seasonal demand based on last year's numbers

After: Predicts demand by analyzing:

- Weather patterns
- Local events
- Social media trends
- Historical sales data
- Customer preferences

The impact? *"We went from hoping we had enough beer for big events to knowing exactly what we'll need,"* Dwight shared. Their accuracy improved from 82% to 98%, cutting waste by 12%.

Automated Decision Support: AI as Your Business Advisor

Think of this as having an incredibly smart assistant who remembers every business decision and its outcome. These systems don't just provide data - they learn what works and offer specific recommendations.

Black River uses this technology to:

- Adjust brewing schedules based on predicted demand
- Optimize ingredient ordering
- Fine-tune fermentation conditions
- Schedule maintenance before problems occur

"The system doesn't make decisions for us," Eric emphasized. *"It gives us better information to make smarter choices ourselves."*

What This Means for Your Business

These technologies aren't just for big companies. Black River shows how businesses of any size can use them effectively:

- Start with one capability (they began with quality control)
- Build on early successes (added demand prediction next)
- Let your needs guide adoption (focused on what mattered most)
- Measure results at each step (tracked specific improvements)

"The key," Dwight noted, *"is understanding these aren't just fancy technologies - they're tools to solve real business problems."*

MAINTAINING COMPETITIVE ADVANTAGE: THE THREE-TIER APPROACH

I've learned that maintaining an AI advantage requires balancing three critical elements to stay ahead. Let me show you how Black River Brewing mastered this balance to become a regional leader in craft brewing.

1. Capability Building: Strengthening Your Team

Think of this like building a sports team - everyone needs to keep improving to stay competitive.

Black River's approach includes:

- Monthly AI skill workshops
- Hands-on practice with new features
- Real-world problem solving
- Peer teaching opportunities
- Cross-department learning
- Production teams learn from sales
- Sales learns from quality control
- Everyone shares insights
- Innovation incentives
- Rewards for process improvements
- Recognition for creative solutions
- Time allocated for experimentation

"We discovered that our best ideas often come from unexpected places," Dwight noted. *"A delivery driver spotted a pattern that led to one of our biggest efficiency improvements."*

2. Process Evolution: Smart Changes at the Right Pace

Success comes from knowing when to push forward and when to let things settle.

RICHARD RICE

Their systematic approach:

- Weekly quick wins review
- What's working well?
- What needs adjustment?
- Where are the opportunities?
- Monthly efficiency checks
- Compare metrics to goals
- Identify bottlenecks
- Plan improvements
- Quarterly adaptation reviews
- Evaluate new technologies
- Update procedures
- Refine training methods

3. Market Positioning: Staying Ahead of Changes

It's not just about keeping up - it's about leading the way.

Black River maintains their edge through the following:

- Customer feedback loops
- Regular surveys
- Social media monitoring
- Direct conversations
- Pattern analysis
- Competitive intelligence
- Industry trend tracking
- Technology adoption monitoring
- Market opportunity analysis
- Strategic planning
- Quarterly strategy reviews
- Innovation road map updates
- Resource allocation adjustments

Results of this three-pillar approach:

- 31% increase in operational efficiency
- 4% reduction in quality variations
- 28% improvement in market responsiveness
- Leading position in regional craft brewing

"The key is balancing innovation with stability," Eric shared. *"We want to stay ahead without disrupting what's working well. This framework helps us make smart choices about when to push forward and when to optimize what we have."*

The lesson? Success isn't about chasing every new technology - it's about making strategic choices that keep you ahead of your competition while maintaining operational excellence.

COMMON FUTURE-PROOFING CHALLENGES

Technology Overwhelm

- **Challenge:** Too many new options
- **Solution:** Strategic evaluation framework

Team Adaptation

- **Challenge:** Keeping skills current
- **Solution:** Continuous learning program

Process Evolution

- **Challenge:** Balancing change and stability
- **Solution:** Systematic review cycles

Market Position

- **Challenge:** Maintaining a competitive advantage
- **Solution:** Regular strategic assessment

Tool #15 - Future-Proof Your AI: The Evolution Road Map Template

To help you stay ahead of the curve and systematically advance your AI capabilities, I've created the Evolution Road Map Template, which helps you plan the strategic evolution of your AI capabilities. This long-term planning framework has helped organizations maintain a 12-18-month competitive advantage while reducing technology obsolescence risk by 70%. Instead of constantly playing catch-up, you'll have a clear path to staying ahead in your AI journey.

FINAL THOUGHTS

As I write this final chapter, I'm reminded of a conversation I had with Craig from TRAXX Flooring. "*Rich,*" he said, "*implementing AI wasn't just about improving our business—it was about re-imagining what's possible.*"

That's my challenge to you: Don't just implement AI; use it to re-imagine your business's possibilities. This book's frameworks, tools, and strategies are not just about keeping up with change; they're about leading it.

Your journey with AI is just beginning. The future is not something that happens to us; it's something we create. Through thoughtful implementation, continuous measurement, and strategic evolution, you can ensure your business survives and thrives in the AI-driven future.

Remember what Tom from EndUp Furniture shared: "*The greatest value of our AI implementation wasn't in what it helped us accomplish—it was in what it helped us become.*"

The question isn't whether AI will transform your industry but whether you'll lead or follow that transformation.

The AI GROWTH Code is now in your hands.

The path is clear.

The future is waiting.

What's your next move?

Part Three: Executing the Code

ABOUT THE AUTHOR

Richard Rice is the AI Strategist business leaders trust to turn hard-to-understand technology into bottom-line results. Through practical AI implementations, he has helped over 120 companies achieve an average profit increase of 13%. As the founder of The AI Pros Agency and creator of the AI Operating System™, he excels at simplifying complex concepts.

A natural problem-solver with a global perspective (from his early years in Germany), Richard earned his AI expertise by tackling urgent business challenges like reducing support queues, eliminating payment delays, and preventing quality control issues. His hands-on approach and marketing technology background make him a sought-after advisor who reminds everyone that AI is "a tool, not an easy button."

When not simplifying AI for business leaders, Richard can be found in beautiful Greenville, South Carolina, with his "smokin' hot wife." He serves as the chief ball thrower for their four dogs and reluctant cat bed for their 32-pound cat.

https://richardrice.me/

www.ingramcontent.com/pod-product-compliance
Lightning Source LLC
Chambersburg PA
CBHW040752220326
41597CB00029BA/4736